Curiosity

Farmers Finding Freedom

John King

Advanced Praise for Curiosity – Farmers Finding Freedom

While still enjoying 'Curiosity - Farmers Discovering What Works' on a regular basis by being able to reread individual essays (or listen to the audiobook version) when specific points, strategies or wisdom shared by a diverse range of farmers, often it is also reading chapters by many I have had the pleasure of crossing paths with and admire for how they have addressed their resource concerns, in itself is a great resource and inspiration.

To build upon this vast coalescence of knowledge with 'Curiosity - Farmers Finding Freedom' will further add a great quantity of thought-provoking ideas, advice and wisdom that can be applied to many facets of farm management and life. Curiosity, as defined in this book, will aid the reader to innovate most aspects of their enterprises or have the freedom and confidence that comes from knowing you're on the right path with your own journey.

It is great to see the robust conversations had among people at Curious Cockie farm visits or other field days facilitated by John and his thoughtful questions fastidiously compiled into book form for many others to enjoy.

Duncan Humm ~ Founder and Editor at NZ Farming ~ Mt Somers, Canterbury, New Zealand

John King has captured the very meaning of the word "Curiosity" in his book "Curiosity - Farmers Finding Freedom" The very definition of Curiosity is the state of being curious: inquisitive, wondering, ready to poke around and figure something out.

John has poked around many, many New Zealand farmers over the years teaching, training and mentoring them through the change required to become profitable stewards of the land. I particularly loved the Eco- profit Observations that summarises each case study. I highly recommend this book to anyone curious to know what's possible if one applies themselves.

Grahame Rees ~ Director ~ KLR Marketing ~ Bathurst, NSW, Australia

Regarding this book, I must read it in the daytime because if I read at night I can't sleep, over-stimulated by the creativity and openness of these farmers to find their model, based on their values and skill sets. Also great to see completely opposing systems working with many realizing the conventional model doesn't have farmers' best interest. These stories are great for young and old, newbies and oldies. And yes many terms about New Zealand farming I didn't understand, but it didn't take away from the read.

Brian Maloney ~ Ferme Brylee ~ Thurso, Quebec, Canada

About the Author.

John King is an adult educator who combines observation and critical thinking to help farmers question assumptions and practices that seem to make their lives easier but work against their long term interests.

John's pragmatism comes from being raised at "The Bend" family farm near Winton, Southland, with farming ancestors going back centuries. He spent several years travelling and working in the USA and UK before returning to New Zealand to graduate with a M. Agr, Sc. in Adult and Rural Development from Lincoln University, Canterbury, and a Diploma in Adult Teaching and Learning from University of Canterbury, Christchurch.

He met Allan Savory, founder of Holistic Management in 1994 and since 2001 began training farmers in New Zealand and Australia and mentoring others around the globe. Exposure to techniques and perspectives through this network underlie many of the ideas and insights he presents.

As an independent facilitator, John organised 17 Red Meat Profit Partnership action groups throughout New Zealand during 2018-2020, some 8% of the network. When Beef and Lamb NZ pulled all funding in 2023 John had 13 action groups operating, 25 % of the network.

From there John founded the nationwide Curious Cockie Club network after farmers found his approach to facilitation provided more flexibility than traditional production focused farm discussion groups. Many stories here are from property visits by these groups.

John continues to write and publish opinions and insights on industry issues and a number of those articles are presented here.

John lives in Christchurch, New Zealand.

For my father Warwick

Acknowledgements.

This book couldn't be completed without the cooperation of the following farmers and professionals; Pita Alexander, Shane Armstrong, Adrian and Pauline Ball, Peter and Robin Berryman and Danny Henman and Liz Berryman, Tony and Angela Blunt, Jeff Broomfield, Colin and Keryn Brown, Hamish Browne and Robyn Budd, Andy Bunny, Forbes and Angus Cameron, Hugh Campbell, Jeremy Casey and Kim Solly, Tim Chamberlain, Guy and Ashley Coats, Louise Coats, Peter Coles, Geoff and Barb Croker, Greg Crombie and Siobhan Wilson, Blair and Penelope Drysdale, Paul and Prue Ensor, Adrian Frei, Jamie and Nicky Gaddum and Robert Hesselsteiner, Bill and Sue Garland, Neil and Glenda Gray, Robin and Lois Greer, Siobhan Griffin, James Halford, Tom Hargreaves, Tim Hawke and Murray Weaver, Russell and Charlotte Heald, Henry Hendricks, Peter and Christine Johnstone, Simon Lee, Travis Leslie, Bryce and Aneta Lupton, Michael Magiera and Smita Apte, Max and Richard Nicol, Dennis Nuiewkoop, Peter and Mandy Paterson, Angela Payne, Vince Payne, Brett Petersen, Will Perriam-Lampp, Ross and Karen Potter, Mark, Judy and Keeley Pudney, Michael and Katherine Reilly, Cliff Shearer, Kenneth and Rachel Short, Roger Slee, Graham Smith, Karl Steiner, Andrew Tabaner, Jeremy Talbot, Richard Thorp, John Vosper, Mark Warren, Daniel Wheeler and Donna Hicks, Willie White.

I would also like to thank proofreader Jon Morgan and acknowledge media outlets Stuff, New Zealand Farmers Weekly, as well as Holistic Management International for pre-publishing stories here.

I am also grateful to the network of Curious Cockies who continue to join me in visiting properties around New Zealand to discover what is and is not working for local farmers.

1st printing

ISBN:
978-1-99-118876-2 (pbk)
978-1-99-118877-9 (Kindle)

Cover photo: John King
Author photo: Tony Benny

Also by the same author:

Curiosity – Farmers Discovering What Works

Contents

Location of New Zealand Farmers

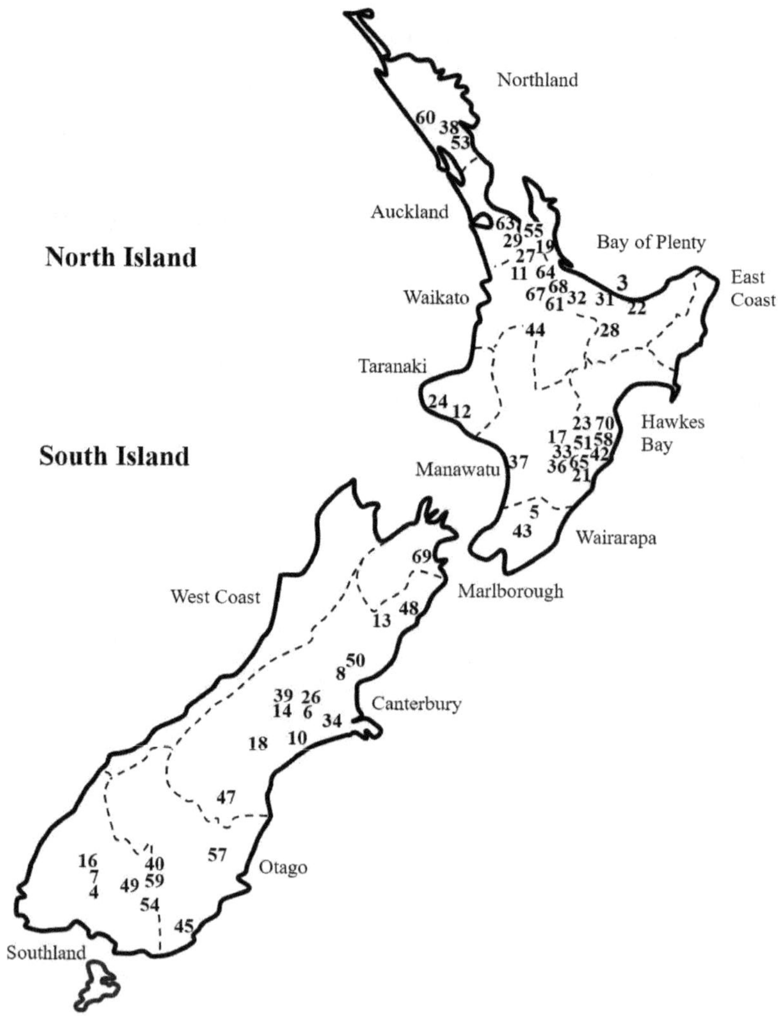

1. What is Curiosity?

We cannot teach people anything; we can only help them discover it within themselves. – Galileo Galilei

Many farmers are driven by curiosity to learn more. There's no cure for curiosity but it can be hampered by a shortage of time or resources, or by the ready-made shortcuts of past experiences, or by community attitudes to what is, or is not, appropriate. What triggers curiosity that leads to farmer actions and observations are the subjects of Curiosity books.

The stories here are largely personal accounts of what works for New Zealand farming operations. Curiosity is not the wonder-like experience of childhood but is a multi-layered journey of advances and setbacks, upending beliefs and prejudices, recycling various emotions while juggling social, business and environmental relationships. Many of these stories challenge the industry narrative because ready-made solutions do not live up to their promise, leaving many farmers frustrated and lonely.

Curiosity is lust of the mind according to English philosopher Thomas Hobbes and often requires a sense of courage. That's where freedom emerges. Freedom lies in being bold, says American rural poet Robert Frost. And yet many never go there because as educators like Sir Ken Robinson state, failure has become so stigmatised. Freedom to fail is what every farmer should strive for because in many instances it represents growth from the prejudices and pressures conforming to convention.

And there lies the challenge. In his 2017 book, Call of the Reed Warbler, Charles Massy says most farmers are unaware they are captured, or that they exist or operate within a particular view of how the world works. They proudly show fenced-off riparian corridors, but to view these they travel through sprayed-out and bare or overgrazed cropping and pasture paddocks (p.410).

And then they have an epiphany, a moment that truly challenges what they believe to be true. Whether it's a singular disaster like a flood, drought, relationship breakup, business failure, accident, illness or death, or a cascade of events gnawing at the foundations of their beliefs, such as inconsistent results, remedies worse than diseases, prescriptions constantly treating symptoms instead of addressing causes, being exposed to inflation, and other circumstances challenging their beliefs by revealing a new reality.

Many stop there, but others reach beyond established friendships and networks to new people and find there's a different way. And this is where the real battle lies.

Many farmers struggle against the industry's own hype. The only celebrated successes are those which fit the industrial model and those engaging in different practices are often shunned. How often do we see organic farmers winning environment awards in New Zealand? It seldom happens because they don't buy products from environment award sponsors.

Don't believe me? If a farmer drops their fertiliser purchases 70% by applying soluble products as foliars and maintains production, would their fertiliser company ever nominate them for a farming environment award? Never. I rest my case.

How many farm environment awards have ever recognised what New Zealand agroecologist Peter Sears discovered in the 1940s, how to use livestock to enhance productive capacities of landscapes? His research highlighted how livestock lifted or maintained fertility by camping overnight on smaller areas than they grazed during the day. This simple technique saves thousands in

Effect of Day and Night Grazing on Annual Pasture Composition and Production.

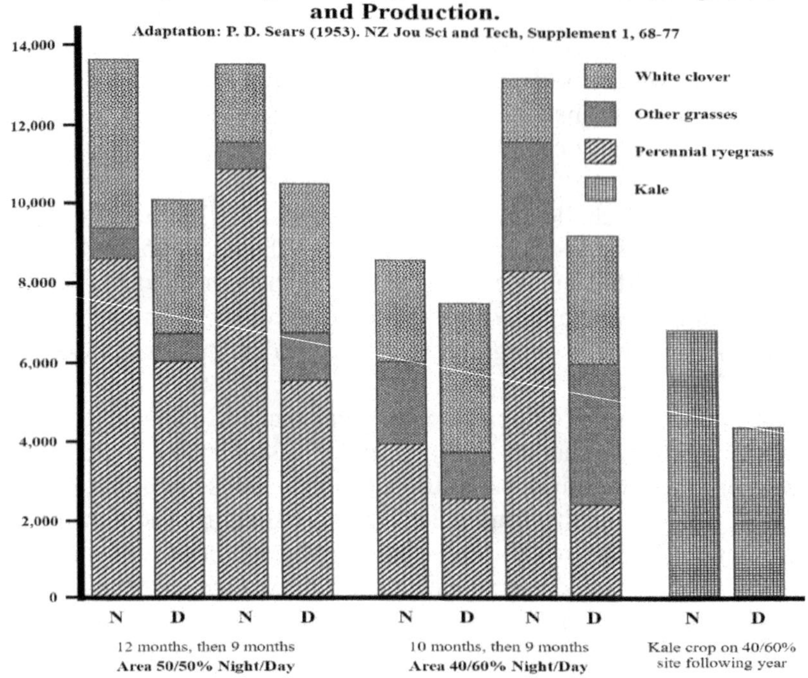

fertiliser bills but it's never mentioned in media or taught by institutions because farming can't exist without purchased inputs.

This is how industrial agriculture successfully negates livestock's role in saving the planet. It fails to incorporate additional services livestock provide, other than grazing to enhance landscape function when planning their movements across properties. Commercial expediency compromises farming's role in building climate and community resilience.

Entrenched attitudes towards alternative ideas is a huge problem in farming. Tribalism drummed up by farming organisations and rural leaders isolate many farmers when industry promises fail to materialise. What level of support can industry agencies deliver to combat community bullying and intolerance when they are staffed by professionals who themselves are limited in their exposure to alternatives? How many struggling farmers are advised to continue what they've been doing but get a $500,000 loan to start again?

Where do farmers start when they're struggling? As Milton sheep farmer Rick Cameron states, freedom to fail comes from the drudgery of numbers. If you're not prepared to observe, monitor and record, fear rather than fun drives your choices. Success comes from getting started and as there is very little industry advice to look beyond maximising production, these stories explore farmers' perspectives and present a library of insights from experiences for those looking for something else.

For readers of my previous book, Curiosity: Farmers Discovering What Works, much of that book focused on the first link in the Solar Chain of Production. That link is essential for optimal harvesting, packaging and selling of second-hand sunbeams to enlightened consumers. You'll find this book delves deeper into the other two links.

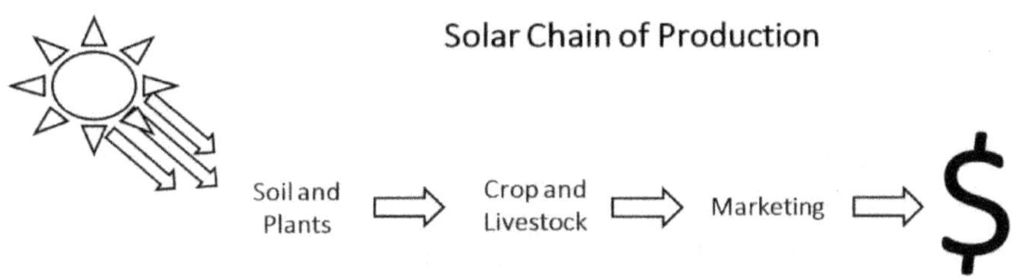

Adaptation by Graeme Hand: Savory, A. (2016). Holistic Management: A Commonsense Revolution to Restore our Environment

If by the end of this book you don't know which of these three links is the bottleneck in your business, it could be you…

2. Valuing Rainfall Efficiency

About 66 percent of a healthy soil is just space, air — nothing — and that creates massive capacity for the ***sponge*** *(my emphasis) to hold water. It allows water to infiltrate and be retained and made available over time. It's really the "nothing" that we add to soil which creates its health and viability...* Walter Jehne, Australian soil microbiologist and climate scientist. In: Restoring Soil via Carbon Sponge Support. April 2019. Acres USA.

Is there a single irrigation company showing farming clients what soil surfaces must look like to maximise water absorption? Nope. Despite regular panic attacks over forecast dry summers, does anybody look beyond the nozzle?

If I ask farmers if water is essential in producing food and fibre for their business, all hands go up. When I ask if they can state rainfall infiltration rates across their property, all hands come down.

While industry focuses on Farm Environment and Freshwater Plans (FEP), does anybody during those workshops remind farmers soil is the largest reservoir of fresh water? Could that be of value?

How often do farmers brag about efficiencies of new tractors or other equipment? When buying an asset, farmers aren't just interested in its colour or horsepower. With current inflation they're also keen on how well it performs.

Compare that to purchasing or leasing land. Is rainfall infiltration and its relationship to kgDM produced ever mentioned when talking about production efficiency? Moisture is nature's lubricant greasing the powertrain driving every farmer's business. As farmers discover, having a perfect soil test counts for nothing if it hasn't rained for weeks.

Farmers carry spades to dig holes and check soils for themselves instead of accepting reports on paper – Challenging Farming Culture.

Even when rain arrives, what production potential do farmers miss if soil is absorbing only 30%, 50% or 70% of what falls? Furthermore, when moisture isn't sinking into soil, its runoff siphons off thousands of dollars of fertiliser with it.

Despite all the talk of precision agriculture, no fertiliser or agrichemical company guarantees their products stay where put. Instead, that's added to farmers' social licence, delivering rising compliance costs borne by them.

Land is the most expensive asset farmers own and yet nobody is curious about rainfall absorption and its relationship to production. When your neighbour rings up and asks, "How much rain did you get last night?" imagine the satisfaction of replying "I got it all mate," knowing your growing period will be longer than next door's.

With industry experts declaring that NZ soils are saturated with carbon, nobody is investigating potential benefits of a Carbon Sponge for communities. Farmers already take a dangerous pollutant, CO_2, and create oxygen and other beneficial products for communities. What potential public good also exists with soil?

Imagine an army of farmers hopping into their tractors, rumbling down to their regional council, knocking on the door declaring, "Do we have a deal for you!"

"We can prove that for 99% of the time and over 99% of the area we farm we can absorb 99% of the rain that falls. We reduce the frequency and severity of droughts and floods in our catchment, thereby minimising wear and tear on community assets. Furthermore, we lower risks for local businesses by reducing boom/bust cycles created by climatic events.

> Covering soil surfaces with dried forage assists with maintaining soil moisture. As moisture collects in rock crevices and cracks, trampling triggers germination of this seed source by ensuring surface contact – How Livestock Save the Planet.

"Our land management is producing a public good of significance. How about a rates rebate of 50%?"

I can't see regional councils rubbing their hands in joy at such a proposal but the story would pressure them to recognise soil as a freshwater reservoir. Farmers could claim the high moral ground by using the term "Carbon Sponge", a label far more descriptive for public insight than FEP.

Would Farm Environment Plans be more relevant to farmers if they led to greater soil storage of rainwater? There's an ecosystem of consultants to help farmers with their environmental compliance. The problem is, compliance embraces a monoculture of entrenched prescriptions, ringfenced from alternative ideas and railroaded by government funding. Soil surfaces never get a mention.

For example, take DoC. Their only tool for tackling climate is fencing land from disturbance. In brittle climates this tool produces bare soil and increases runoff. Elsewhere it creates high fuel loads for wildfires and eroding stream banks choking waterways with gravel and tinder dry, herbicided woody weeds. None of these outcomes contributes to a functioning Carbon Sponge for storing water. Interestingly, any ecological succession starts by changing the soil surface.

Will MPI sustainability advisers be any different? Probably not. If professionals don't know what to look for to increase infiltration rates, then what chances do farmers have?

Stockman Grass Farmer editor Allan Nation made this observation to rise above all agendas. He stated yields are largely a function of soil organic matter, largely determined by how long land is in pasture, which is largely determined by grazing management. For him, soil organic matter simply reduced climate risks and created opportunities for profit. The deeper soil organic matter goes, the bigger the buffer.

Recently I heard commentators claiming New Zealand farmers are 20 years behind North American ranchers in their grazing management. Could that reflect our fascination with post-indicators like biomass compared to pre-indicators like rainfall infiltration? As mechanics observe, it doesn't matter how powerful your engine is if you're using the wrong dipstick.

Eco-profit Indicators

- No irrigation company or regional council shows farmers what soil surfaces must look like to maximise water absorption.
- Moisture is nature's lubricant greasing the powertrain driving every farmer's business, yet rainfall infiltration and its relationship to biomass produced is never mentioned.
- What production potential do farmers miss if soil is absorbing only 30%, 50% or 70% of rainfall or irrigation?
- No fertiliser or agrichemical company guarantees their products stay where they're put, yet it's the farmer who bears that cost from surface runoff or leaching.
- Nobody is investigating potential benefits of a Carbon Sponge for communities.

- Farmers could claim the high moral ground by using the term "Carbon Sponge," a label far more descriptive for public insight than FEP.
- Soil organic matter, or Carbon Sponge reduces climate risks and creates opportunities for profit.

3. Brett Petersen - Where to Start?

It was year after year of 30% potassic super without any pasture improvement that got Brett Petersen to ask questions. He threatened to do something, but what? By now he'd burned years of university notes on horticulture, agriculture and farm management because those recipes were not producing results.

In 1980, as a Department of Māori Affairs supervisor developing kiwifruit orchards and overseeing various farms, he noticed how spraying glyphosate under vines created soil surfaces covered with moss. He thought, "For 400 million years, moss secreted acid onto rock to create soil. Now, it takes us only two years to turn soil into rock."

He began questioning the integrity and sense of many practices. For example, girdling kiwifruit vines to restrict sap flow to and from roots. Some 60% of soluble compounds produced in leaves transfers to roots, of which 50% exudes into soil to feed microbes. In return, microbes solubilise nutrients for plants. Girdling destroyed this synergistic relationship.

Consequently, orchard soils degraded and compacted with severely compromised roots. Some orchards girdled up to four times a season, leading to poor vine growth years later. All short-term financial gains from girdling stripped longer-term profits, something not revealed to unsuspecting new market entrants. Pasture burnout from overgrazing is similar.

In the 2000s, at an Arden Anderson course, he heard information that resonated with his experiences. He was motivated to fly to Queensland to hear Graeme Sait. These events made him realise university education had not taught him the full facts, particularly about soil fertility. He recalls being unable to interpret soil tests, leaving recommendations to industry professionals who knew no better.

About this time, he was developing dairy farms for Māori trusts. After spraying and oversowing with pasture and superphosphate, he noticed former scrub and blackberry patches became clean pasture, but areas of clean pasture grew many weeds. The farms were not improving. It was here Brett took the plunge and tried several different companies' products and systems.

He joined eCogent and found a network where 90% of clients were on an alternative fertiliser programme, an outcome they figured out for themselves. eCogent focused on measuring inputs and outputs in cents/kgDM. This allowed farmers and consultants to identify what drove profit, rather than chasing production and lifting costs.

eCogent's system focused on balancing daily feed supply and demand. The cheapest feed is that eaten, followed by that made into supplements, then bought feed. Therefore, making supplement was by accident, not design. Many found milking a quarter of their herd over winter helped balance feed slumps and bumps, and maximised profit.

Brett supervised one Māori trust farm of similar size to the Lincoln University Dairy Farm that squeezed out 850kgMS/ha on stony river flats. While, LUDF's best ROI was 6.4% with twice the production, Brett's property averaged 8.6% and topped out in 2008 with a high of 12.8% ROI. At that time, the New Zealand average was 2-3%.

Brett puts this down to having the property's chequebook himself, allowing him to focus on profit rather than production. It highlighted eCogent's processes to focus on expenses that made money. His bookkeeping was so meticulous, accountancy fees rose more than 10-fold when he left.

Brett has experienced many benefits from changing soil fertility inputs in the move away from conventional industrial farming.

> An important observation is cows start eating pasture once inside gateways instead of bouncing off paddock fences looking for something – Peter and Mandy Paterson.

If shifting to a new fertiliser regime, Brett has observed livestock behaviour changes before production improves. Firstly, anxiety reduces and animals become more relaxed. Instead of running up to and through a paddock gate and charging to the back fence to look for something suitable to eat, they amble up and then calmly put their heads down inside the gate. This initially creates a jam as the rest push past to get in.

Then their milk quality and production increases, particularly fats, along with body condition. Other signs of improving animal health include higher calving rates, longevity of cows, less metabolic disease and retained membranes, even less intestinal worms. Essentially, animal health costs decline, allowing farmers to pocket more money.

Pest and diseases that thrive in acidic conditions from modern fertilisers also disappear. He's observed facial eczema, army worm, black beetle, grass grub and porina just fade away.

> Drought was visiting again and he calculated lifting soil carbon by 1% annually would increase water-holding by 100mm over four years, allowing grass to grow for another month – Tony and Angela Blunt.

After abusing soils for 60-80 years, it's no surprise they take longer to respond than cows and pasture. Always build soil structure before adding biology otherwise it won't thrive. Once the calcium:magnesium ratio gets into balance things change. While root depth can change in one or two years, it can take five years to see noticeable differences. To fully detox soil can take 20 years. Pace of change increases if soil is microbially alive than if chemically dead.

As soil structure improves, root depth deepens and soil aggregates enlarge, which advances both drainage and water storage. That lengthens growing periods between rainfall events, meaning pastures last longer into dry spells and exit dry and wet spells earlier.

Many farmers understand this pathway involves liming but Brett warns not to confuse calcium, lime and pH. Citing US soil scientist Dr William Albrecht's demonstration of drilling soybeans with equal amounts of calcium as calcium hydroxide (alkaline), calcium nitrate (neutral) or calcium chloride (acidic), produced identical crop yields independent of soil pH. All yields were superior to soybeans grown without calcium.

Brett ended up joining Kiwi Fertiliser where his many years with orchards and farming made him a versatile consultant. His experience highlights that if things aren't working and you're looking for answers, you've got to push your own boundaries and go beyond the people you already know.

Eco-profit Observations

- Questioned university teachings as they were not producing results.
- Spraying glyphosate turns soil into rock.
- Girdling kiwifruit vines to restrict sap flow to and from roots produced compacted orchard soils. Overgrazing pasture does the same thing.

- Inability to interpret soil tests left him vulnerable to industry recommendations.
- Inconsistent development outcomes on dairy farms made him look beyond industry professionals.
- Measuring inputs and outputs in cents/kgDM allows farmers and consultants to identify what drives profit.
- Cheapest feed is that eaten, followed by supplements, then bought feed.
- Milking quarter of a dairy herd over winter balances feed demand and supply which assists in maximising profit.
- Using these techniques doubled the profit of local farms versus the Lincoln University Dairy Farm run on industry-best advice.
- Upon leaving his role, accountancy fees rose more than 10-fold.
- When changing production systems, livestock behaviour changes before production improves.
- Production improvements include: Milk quality and volume, particularly fats, higher calving rates and longevity of cows, less metabolic disease and parasitism.
- Pest and diseases that thrive in acidic conditions also disappear.
- Root depth can change in one or two years; it can take five years to see noticeable differences. To fully detox soil can take 20 years.
- As soil structure improves, root depth deepens and soil aggregates enlarge, which advances both drainage and water storage.
- This path often involves liming but do not confuse calcium, lime and pH as crop yields demonstrate.

4. Jeff Broomfield - Never Buy Better

Working on Briscoes' Invercargill shop floor in the early 1980s, gave Jeff Broomfield a front seat view of business strategy. The practice of dropping staff and lifting revenue became the business model for two of New Zealand's most iconic brands, Briscoes and The Warehouse. It highlighted how running at small margins with higher volumes creates more turnover. So how does the You'll Never Buy Better slogan relate to farming?

By the mid-eighties Jeff was farming at home, at Otautau. New Zealand was in a recession and he cranked up production from 100kgs of superphosphate every second year to 250kgs every year, plus urea. He grew a lot of grass and lifted production 400%, but profit remained static. Someone was making money and it wasn't him. Furthermore, problems were emerging on-farm.

Crop health tanked within three years. Soil compacted and became blocky, often requiring power harrowing twice after rolling and ploughing because clods didn't break down. While crops grew, they needed more insecticides and fungicides to deal with pests.

Then pastures ran out of grunt from being kept constantly vegetative from topping and silage harvesting. Instead of Moata ryegrass pastures lasting seven years, they barely survived two. Making balage simply pickled seeds. Without hay, paddocks had little chance of reseeding themselves, reducing pasture longevity.

Livestock problems took 10 years to appear. By then, sheep had bearings and metabolic problems, including milk fever and sleepy sickness, and even the odd fungal outbreak and circling disease.

Jeff can't remember what tipped him away from synthetics, probably a column by advisers Lloyd Wilson or Ron Halford in late 1990s. But he recalls getting good results applying fish. He started messing around with compost-based teas and foliars.

New Zealand's soil health debate intensified after 2000. With Albrecht's model gaining a foothold, traditional fertiliser companies challenged claims of better production. Diving into Neal Kinsey workshops with a background in biological products, Jeff brought a different view on how Albrecht's processes worked. By now Jeff was associated with Healthy Soils and their investment in a substantial research programme.

He left Healthy Soils only to return 18 months later to pick up the intellectual property when it fell into receivership. Then he partnered with Methven cropping farmer and topdressing pilot Don Hart from Top Soils.

> Then the scientists put soil lysimeters on both sides of the fence to measure nitrogen leaching. Within two years scientists removed them – Peter Berryman

While Albrecht's model remains a base for advice, Jeff uses biologically friendlier products. Many of his insights involve humates, a compound formed by aerobic decomposing of organic matter such as cellulose, proteins and starches. When pressure squeezes out oxygen it becomes lignite. Interestingly, lignite burns, humates don't. Jeff wrote the humate protocol standard for New Zealand. Now all humates are at a good standard.

Mixing with innovators beyond New Zealand's corporatised ag advice community developed new perspectives to challenge soil chemistry. Jeff learned that the half-life utilisation of elements is determined with conventional fertilisers, not biological ones or combinations. When formulas are changed, so does utilisation and efficiency.

For example, humates with urea lift pasture production by 20-30%, mostly from reducing nitrogen leaching by 60%. How can a Briscoes shop floor ex-employee see value here, but not university graduates?

Another example is phosphorous. Of the 90kg of phosphorus per tonne of superphosphate, plants readily access 10kg. With 140kg of phosphorous in a tonne of calcium phosphate, plants access 126kg. This means that, at current prices, the cost of available phosphorous in super equates to $47/kgP compared to $5.40/kgP with calcium phosphate, or one tonne of super equates to 80kg calcium phosphate.

Furthermore, calcium phosphate increases bacteria seven times and fungi five times, snowballing soil's capacity to hold nutrient. So does a soft phosphate rock like guano which boosts dry matter 5-7 times higher than superphosphate coinciding with bacteria lifting 4-7 times and fungi 3-4 times over a nine-year trial.

Soil organisms help in other ways. For example, solubility of sulphur declines as bacteria stop converting it into biological compounds and sulphur dioxide, stabilising its presence on a soil test. That lifts production. All these insights are about soil doing more from less fertiliser.

Additionally, a lack of soil biology renders products such as calcium magnesite useless because magnesium isn't soluble. Prior to 1990, serpentine super was readily used on pastoral farms. It's slow-release magnesium is why livestock health problems took 20 years to emerge when replaced with straight super. Plants use magnesium for photosynthesis and it reduces animal anxiety. Both have an effect on production.

What compounds magnesium issues is how industry professionals use MAF soil Quick Tests. Developed in 1950s, these are accurate for everything except animal health. Why? Quick Tests state figures of 8-11 are good for crop production, but adds only in fine print that animal health requires 28-32. This means a reading of 10 delivers 40% of what livestock need, leading to many animal health problems. But who reads fine print?

> Breeds also have differences with trace element consumption. Texels originally come from copper-deficient islands whereas Finn derives from places with higher copper levels – Daniel

Back at Otautau, Jeff's property is all cut-and-carry. Yet pasture longevity is pushing beyond 15 years because CEC and soil nutrition is functioning properly, despite little recycling of pasture litter.

So, how does Briscoes' business principle provide a lesson for farmers? Applying smaller amounts of product with greater utilisation lowers costs and improves nutrient efficiencies. That reduces pollution, lifts production and increases profits. Jeff assures us, "You'll Never Buy Better".

Eco-profit Observations

- Growing lots of grass lifted production 400% but profit remained static.
- Soil compacted and became blocky, often requiring more cultivation.
- Crops grew, but they required more insecticides and fungicides to deal with pests.
- Balage pickled seeds and without hay, paddocks didn't reseed themselves, reducing pasture longevity.
- Livestock problems took 10 years to appear, including bearings, milk fever, sleepy sickness, the odd fungal outbreak and circling disease.

- He started getting good results applying fish. Then he applied compost-based teas and foliars.
- While Albrecht's model remains a base for advice, Jeff uses more biologically friendly products from his insights with humates. Jeff wrote the humate protocol standard for New Zealand.
- The half-life utilisation of elements are determined with conventional fertilisers, not biological ones or combinations. When formulas are changed, so does utilisation and efficiency.
- Humates with urea lift pasture production by 20-30% and reduce nitrogen leaching by 60%.
- Calcium phosphate increases bacteria seven times and fungi five times, snowballing soil's capacity to hold nutrient.
- Solubility of sulphur declines as bacteria stop converting it into biological compounds and lifts pasture production.
- Reduction of serpentine super and slow-release magnesium is why it took 20 years for animal health problems to emerge.
- Read the fine print. Quick Tests state magnesium figures of 8-11 are good for crop production, but animal health requires 28-32.
- Applying smaller amounts of product with greater utilisation lowers costs and improves nutrient efficiencies.

5. Andy Bunny - Wean off Fertiliser

Andy Bunny's journey started when he dumped 450kg/ha of superphosphate on a block to lift soil Olsen P of 8. But production went backwards and Olsen P didn't move.

Farming is challenging enough in Wairarapa, wet cold winters, hot dry summers, and that relentless wind. Andy's 440ha hill country farm near Masterton, Blackrock, was settled by his great-grandparents in 1890 and now includes 230ha leased from his parents next door plus a recent purchase of 150ha. He runs 3000 ewes, 1000 hoggets, 140 cows and puts 50 heifers to the bull.

Andy didn't know what he was looking for but conventional advice wasn't working for him. Being in discussion groups where farmers shared production and financial figures, he felt many were not getting ahead despite advice being offered. With limited fertiliser knowledge he began learning how it worked.

YouTube became his library. There he found people talking a new language: Livestock are good for landscapes and fertiliser creates more problems than grazing animals. Speakers grabbing his attention included Dr Christine Jones, Gabe Brown, Dr Harley Smith, Elaine Ingham, Nicole Masters and in recent years Jaime Elizondo. Social media and regen discussion groups created opportunities to ask questions and bounce ideas.

> Imagine buying a dairy farm and then having 10-15 cows going down with milk fever each day. You end the season losing 10% of the herd. What would you do next? – Mark, Judy and Keeley Pudney.

As Andy delved into the regenerative path, he found synthetic fertiliser significantly impacted soil biology and health by removing the abilities of fungi and bacteria to function and build organic matter. It also tied up various trace elements, causing animal health problems and hindering performance.

Many pasture and soil science trials use sites with decades of fertiliser inputs so their results become irrelevant when running a natural farming system with biology functioning properly. From here he began researching how to wean his farm off synthetic fertiliser and chemicals and how to reduce animal health treatments.

Going cold turkey turns farms yellow as they lack biology to replace pasture reliance on solid fertiliser. Since 2017 he's used minute amounts of high-analysis fertiliser. Andy believes in three ways to wean off solid fertiliser. The first is to drop rates by half for three years in a row, the second is to use foliar sprays to feed plants and biology directly or, the third way, granules (fine particle fertiliser).

Liquid foliar fertilisers deliver nutrients more efficiently than granules. Granule urea volatilises but adding liquid nitrogen to molasses, seaweed, fish fertiliser, fulvic acid or humates binds it to carbon. This acts as a chelator by preventing it attaching to, and leaching, magnesium and calcium.

As Andy points out, fertilisers like superphosphate don't go into soil. Instead, they stay on surfaces locking up calcium, magnesium and zinc which all get washed down hills when it rains. He estimates his natural phosphate losses are 1.8kg/ha/yr while rain delivers 2kg/ha/yr.

Also, over-liming ties up other minerals like copper. Andy concentrates 25% of his animal health budget on four minerals, copper, iodine, selenium and cobalt. Low copper leads to parasite infections in lambs and ewes so rock salt, magnesium and copper blocks are available in every paddock.

He's found summer crops too expensive. When looking at pasture renewal he doesn't waste money on 20-species mixes. He's only used them to figure out which species grow well. For him those are chicory, plantain and red and white clover. If pasture renewal is needed, he mainly sows these. He doesn't sow any grasses as the seed bank is already there.

Over the years, he's tried different grazing methods but believes his property grows better pasture by grazing hard throughout the whole year. Elizondo's Total Graze was the spark he needed. Rotational grazing and Total Graze are different. Total Graze is eating all plants and stems. Grazing pasture short with a long recovery period is key to ensure photosynthesis stimulates soil biology and makes nutrients available to plants. This improves pasture quality and quantity.

Key to changing is bigger mobs and non-selective grazing. It works best shifting large mobs four times per day, but Andy mostly shifts daily. He has no qualms putting large mobs together as cattle become quieter and easier to shift. While he splits mobs for tupping and set stocks for lambing, once the ram has been out for two cycles, he can put all ewes together. The days of little mobs around the farm or opening gates up when it's dry are gone.

Hard grazing with dense mobs either consumes or tramples everything which thickens swards by germinating more seedlings, ensuring pasture quality from the soil surface up. He's found higher post-grazing residuals open pastures up to bare soil and weeds. When grazing with sheep, any paddocks with stems lose quality and don't bounce well after rain.

It's all about making money with good pastures, no need for fancy pasture species or mixes. Work with what you have – Bryce Lupton.

The cattle mob either rotates on their own or follows behind ewes to prevent injuring sheep or restricting ewe access to water troughs. It also cleans up all roughage. Any pasture excess is defer-grazed, sometimes up to 12 months. It's the best way to lift soil humus.

With low strong wool prices and rising costs, Andy moved to easy-care Wiltshires in 2021 and is amazed at how tough his 2500 Wiltshires are. Within a single season they outperformed Romneys bred over five generations by hanging on to body condition. He initially worried about them blowing bearings at lambing. Nobody has cracked research around bearings but Andy thinks magnesium is influential.

While still spending a lot of time talking and listening to people, Andy feels he has come a long way in moving away from a chemically driven system on his farm to a natural system with low costs. With good stock genetics, soil biology and grazing management, soil should look after itself.

Eco-profit Observations

- Dumping 450kg/ha of superphosphate to lift Olsen P of 8 resulted in production dropping and Olsen P didn't move.
- Synthetic fertiliser significantly impacts soil biology which ties up various trace elements causing animal health problems and hindering performance.
- To wean off solid fertiliser, first is to drop rates by half for three years in a row, second is to use foliar sprays to feed plants and biology directly or, third, granules (fine particle fertiliser).
- Liquid foliar fertilisers deliver nutrients more efficiently than granules.
- Estimates natural phosphate losses are 1.8kg/ha/yr while rain delivers 2kg/ha/yr.
- Summer crops too expensive. Uses 20-species mixes to figure out which grow well.

- Grazing pasture short with a long recovery period is key to ensure photosynthesis stimulates soil biology.
- Key to changing is bigger mobs and non-selective grazing. It works best shifting large mobs four times a day.
- Within a single season Wiltshires outperformed Romneys bred over five generations by hanging on to body condition.

6. James Halford - Planting the Seed

When changing practices, James Halford asks "How do you know you are doing the right thing? We are the research, not the suppliers as in conventional. I'm trying to educate myself with little one-hectare trials by turning off and on applications."

There's an apprenticeship stage to learning something new, and it's much longer than people expect. A scattergun approach makes it difficult to work out what brings results. Changing systems is frustrating for any farmer on their own, especially when it's not backed by the industry.

James visited New Zealand four times before selling up his UK property in 1997, packing all machinery into containers and settling on a 190ha dry land cropping farm near Methven, Canterbury. His first and second years were commercial disasters. Things were so bad he didn't pay tax for seven years and even got a community services card. He farmed conventionally using cropping practices that meant lots of cultivation.

> Fertiliser requirements are calculated by removing cows so applications target sheep production - cows are to clean up tag – Greg Crombie.

Then around 2018, Don Hart from Top Soils gave him American regenerative farmer Gabe Brown's book Dirt To Soil. It inspired him to start changing practices. While knowing agrichemicals and cultivation drive soil bacterial dominance, common on most cropping farms, he asks what do farmers do with that information?

For James, the first step was to go to a Quorum Sense event. By lunchtime he'd cancelled an order for a new plough and ordered a direct drill. Then a Soil Foodweb course encouraged him to buy his own microscope. Now he spends a lot of time learning about microbes. Figuring out how to work with nature has become a passion as his experiences reveal many agrichemical applications in conventional systems are uneconomic.

With his mortgage paid, he jumped in feet first. However, in hindsight he doesn't recommend this due to crop establishment failures. The best dollar-for-dollar investment on cropping farms occurs at sowing stages because crop establishment is essential. Everything goes downhill if crops can't get out of the ground. It's the hardest puzzle to solve without chemicals. There are many challenges…

Including, how to create a cropping soil with the right fungi:bacteria ratio. Pulling soil samples every year and doing Albrecht cation balancing appears to be working, but James struggles to understand it. Liming is rectifying issues because earthworms are reappearing and soil structure is improving. Penetrometers now reach 400mm before hitting clay. Other solid fertiliser products include dolomite, lime and various trace minerals such as boron and molybdenum.

Nitrogen is a frustrating issue. What works for reducing leaching isn't helpful for starving crops. James is not looking to lower nitrogen rates as crop residue doesn't break down fast enough and robs nitrogen from growing crops. Hence, he doesn't apply humates any more.

He's just completed a one-year trial comparing yields from applying 190kgN/ha as a soil surface dressing versus 130kgN/ha as a standard foliar. Foliar application produced half a tonne more but other trial farmers found the opposite. After a challenging summer, he'll give it another go next season. Nothing appears straightforward.

By using chicken litter, pasture production is reaching nearly 14,000kgDM/ha, as much as they were doing conventionally – Robin and Lois Greer.

Finding uncoated seeds is also a challenge. He now sows bare seed with Australian David Hardwick's biofert composed of EM, for the mother brew, then adds milk and molasses to generate various bacillus bacteria. He'll also add rumen contents and horse dung. He can then create brews that are phosphorous, calcium or mineral heavy by adding lactobacillus to absorb and carry a desired mineral.

His seed drill injects liquids in the furrow for bare seeds. This includes trichoderma, fish, molasses, EM, phosphate fixing bacteria and Contego Turf to deal with the grass grub that affects crops in patches. To inject furrows with biology, the drill pumps 1.5 Bar (22psi) to deliver 100 litres/ha at 10km/hr through a flow control system. No agrichemicals are ever used in the tank. He also drills molasses at one litre/ha in the furrow to feed soil life.

Crop residue causes other dramas than stealing nitrogen. Without soil disturbance and fire, pests are harder to find under a blanket of straw. Slugs are a huge problem, especially grey slugs with autumn sown crops when herbivory easily outcompetes plant growth unlike spring. Incidentally, there are 35 native grey slugs but fortunately all eat dead matter.

He hits slugs early and hard with Ironmax, costing $40-$70/ha. The high up-front expense reduces problems later. His arsenal still includes specific pesticides, as well as herbicides including Roundup. In the autumn of 2024, he redrilled with double the rate of slug bait to get started. Direct-drilling creates a home for slugs by cracking topsoil for slugs to lay eggs. In England, farmers Cambridge-roll soil, forcing slugs to the surface for control with baits.

Ideally, he'd like to promote scarab beetles, centipedes, spiders and others that eat slugs, so he doesn't use insecticide or pyrethroids. He draws on problem insect advice from UK academic Gordon Port, though such research doesn't reflect the scale of commercial farms.

Measuring something can make you change practice. James joined with local and UK farmers interested in fewer agrichemicals. These groups of 5-6 farmers are engaged in on-farm trials, replicating practices and recording data from combine yield monitors for comparison. His local group travels to Australian no-till conferences for further research.

James observes how few problems you have when you don't go looking for them. A friend across the road has a simple conventional system of silage, wheat and barley and does very well. Going down the regenerative path has pushed him, even politically - he's now more left-leaning. While prepared to lose money, he's seeing the light and now breaking even.

Eco-profit Observations

- Experienced commercial failure using conventional cultivation.
- Agrichemicals and cultivation drive soil bacterial dominance.
- Experience reveals many agrichemical applications in conventional systems are uneconomic.
- Crop establishment without agrichemicals is hardest puzzle to solve.
- Lime encourages earthworms.
- Doesn't apply humates as he needs nitrogen to breakdown crop residues while growing crops.
- Foliar application produced half a tonne more grain than surface application, but he could be an outlier.
- Coats uncoated seeds with biology. Can use lactobacillus to carry specific minerals.
- Drill squirts biology and their food into the drill furrow.
- Cambridge-rolling soil forces grey slugs onto surfaces for better slug bait kill.

7. Andrew Taberner - Scent of Success

Some farmers are naturally thrifty when it comes to spending money. Near Te Anau, Westmoor Farming's manager, Andrew Taberner, is a classic example. His curiosity sorting production issues led him down an unplanned track.

With Olsen Ps of 40 and a soil pH of 5.6 on his hill block, it's clear phosphate isn't the limiting factor. Yet local fertiliser reps repeatedly stated, "Just stick with 250kg of 0-9-0 and you'll be right." He rejected that and decided to embark on his own journey to seek solutions. Olsen Ps on his flat land were 10-12 and he needed to lift fertility cheaply. This spurred an intense interest in biological soil conditioners to reduce a reliance on synthetic fertilisers.

> They ended up with Olsen P over 100 and cows doing 290kgMS on twice-a-day milking. Expensive and frustrating – Blair and Penelope Drysdale

Observations backed this decision. Dropping superphosphate and potassium fertilisers resulted in a surprising absence of flystrike with only one case in 1,600 lambs during the 2022/23 summer despite a high fly season. Furthermore, while DAP produced a 15-tonne winter crop but more daggy sheep, commercial fish-based soil conditioners firmed up dung.

But working with biological conditioner start-ups proved challenging. His first winter crop sprayed with fish grew 17 tonne but only five tonne the following season due to a dry summer and a changed formula. That crop turned yellow at six inches high due to nitrogen lock-up.

He began hunting for recipes and resources to do it himself. He found free whole fish carcasses at Sanfords in Bluff. Collecting free stuff is easy, it's having enough time to figure out how to use it that proves troublesome.

Brewing soil conditioners is simple, it's the handling and extraction that's tricky. For example, he made 8,000 litres of salmon and molasses only to find salmon skins clogged up his spraying kit's filter. That was a long afternoon.

So he made a sieve from an old steel wool press lined with hessian sack. Unfortunately, hessian fibres swelled when wet and became impenetrable.

However, showcasing dairy by selling raw milk is a contradiction. After all, the first question a doctor or nurse asks someone with an upset tummy is: Have you drunk raw milk? – Neil and Linda Gray.

Screwing down the lid created so much pressure it split the sides of the press. Back to the drawing board.

He began exploring "fish milkshakes" with ratios 1:1 fish/sugar, 2:3 fish/molasses and 1:2 crays/molasses. All brews used homemade lactobacillus to initiate fermentation. After many trials he found a ratio of four parts minced fish, one part molasses, three parts water and 0.5 parts lactobacillus makes a manageable product with very few solids to remove.

Molasses is better than sugar as it's cheaper and easier to handle. It's also a great preserver as adding ten litres to a 1000 litre tote will keep brews quietly ticking over for 2-3 months. Fish brews take 8-9 months for smells to subside, hence best sited a long way from the house. Once they smell sweeter, they're ready for application.

Andrew also began making lactobacillus from rice and milk with an internet recipe. All his brews start in ten litre buckets as trials. For example, 1kg of rice will sit in four litres of water at 24C for a week before adding its wash to 40 litres of milk for another week. This is scaled up to 20kg of rice in 80 litres of water which is added to 800 litres of whole milk.

After a curd has formed, it's fed to pigs or dogs, (whoever gets there first), and the yellowy coloured serum is transferred to a sealable container with a little molasses for a food source. This is set aside in a cool place away from direct sunlight.

Lactobacillus is a common probiotic known for increasing digestibility. As Andrew reared bobby calves and added a commercial probiotic to their milk at feed times, he became curious about substituting with home-grown lactobacillus instead. Adding 5ml per head with calf milk at each feeding resulted in firmer stools. The smell of ammonia disappearing completely from the calf shed within 36 hours. It cost $1.40/litre to make versus $16/litre for the commercial probiotic, 70 cents/calf versus $8.00/calf.

He also found adding four litres of lactobacillus in water troughs every few days increased dairy heifer weight gain from 500 to 800gms/day reflecting better digestive metabolism due to microbes.

With winter crops, utilisation and satiety is paramount. One pass with 20 litres of seawater, sourced from Te Anau's lobster factory, combined with fish hydrolysate six weeks before grazing, made turnips more palatable. This resulted in contented dairy heifers eating them before eating balage while heifers on untreated turnips ate balage first. When Andrew swapped mobs across, cattle changed their preferences within three days.

Not only can he brew-up biology, he's also mindful of making a home for it. Early pioneers in Maine, US, spread 3 ton/acre of lobster as pasture fertiliser. Endo skeletons are full of chitin which make a great home for biology. A local source is Te Anau's lobster factory. Carcasses are crushed through a mechanical grinder and left outside for fly maggots to eat the flesh, another job downwind of the house. Once dry, it's stored in wool sacks before harrowing into new pasture paddocks before sowing the following season. It's quite compatible with calcium.

When it comes to innovation, Andrew's advice is "Never give up. Expect seven out of ten ideas to fail, meaning there'll be three winners. Try new things, even when neighbours think you're nuts. Look for what is available locally and cheaply. And then observe how any changes affect stock eating habits and health. They are your best indicators."

Eco-profit Observations

- If Olsen P is 40, phosphate is not the limiting factor.
- Focused on biologicals to reduce synthetic fertiliser costs.
- Dropping synthetic fertiliser reduced flystrike. Fish-based soil conditioners firmed up dung.
- Biological conditioner outcomes affected by recipe changes and weather.
- Finding free resources is easier than finding time to work with them.
- Brewing biological conditioners is relatively easy, extraction and handling is more costly.
- Efficient biological conditioners have few solids.
- There are many ways to create fish milkshakes. Molasses is better than sugar.
- Makes own lactobacillus bacteria, great digestion stimulant for calves and increasing liveweight gain in heifers.
- Seawater increases utilisation and satiety of winter crop.
- Ground lobster skeletons are compatible with lime and create homes for soil biology.

8. Tim Hawke and Murray Weaver – Fermenting Farm Fertility

Tucked away from norwesters in the shadow of the Okuku Range stands the original White Rock woolshed built in 1892 as a 16-stand for blade shearing. The photos on the wall show huge pens of sheep outside waiting to be shorn. Originally, White Rock was 44,000 acres, running from Loburn, Canterbury, to the Waipara river 20km away. Lees Valley livestock were hunted over the Okuku range and there was a lookout originally constructed on the woolshed roof to keep an eye out for them.

Tim Hawke's ancestors arrived in 1913 from Claremont, South Canterbury. When Tim took over from his father in 1985, he could name all the families farming up to Rangiora 20km away. This was Corriedale country, but now the district has completely changed and is all dairy cattle and lifestylers.

Like most new farmers, he spent much of his time learning from his parents and their generation. In the late 2000s he joined a Beef and Lamb Performance for Profit group with the aim of lifting production by 20%. His high-performance farming neighbours gave him the tools to step up. If lamb prices had remained constant, profit would have lifted 50%.

He began seeing soil as his greatest asset and wondered how it could advantage livestock further. Traditional fertiliser companies were disappointing – the reps would look at carrying capacity, make a recommendation and then were out the gate. Then a field day at Broomfield hosted by Rob Flynn and Nicole Masters triggered a new thought pattern. After that he started using Rob and Soil Matters and production figures began rising. When previously he struggled to get lambs past 30kg liveweight, now his leasee, Murray Weaver, manages 50kg.

Super didn't grow clover, but lime did as soil pH lifted. He has applied guano and serpentine super but no acidic fertiliser since 2010. He tried fish oil but it got expensive. Then Paul Daly, of EMNZ, introduced him to EM (Effective Micro-organisms) to replace fish oil. It was more versatile; he could spray it on soil, put it in water troughs and drench stock. And he could do all that himself, no more contractors. Production changes were almost immediate with lambing lifting to 175% with 50% off mum to the works. After years of graft and stress, he couldn't believe the ease with which success came.

How simple? Worm drenches strip gut microflora, so with monthly drenching of lambs he now drenches EM 5-10ml/hd afterward as a probiotic to repopulate their gut. Ewes' 10-20ml/hd are done at weaning and pre-lambing. Healthier ewes cycle better and squeeze out more lambs.

He started brewing his own EM, expanding it from 20 litres, plus molasses, up to 200 litres totalling 2000 litres per year. Now he sprays 5 litres/ha on pasture and up to 20 litres/ha on fodder crops. Pastures are sprayed twice a year. He has also started making compost to feed EM and produce fertiliser.

Better animal performance coincided with lower pasture crude protein and fibre but higher dry matter, sugars and ME. The biological side produced healthier grass – Jeremy Casey and Kim Solly.

His brother has a digger and truck which makes getting materials and turning compost over easy. Mineral values are comparable to solid fertiliser so he doesn't need other materials to top it up. As EM is a soup of living biology, it can survive a couple of months before spreading. It can also be applied in the rain, offering tremendous flexibility.

Paul Daly then introduced Tim to Bokashi compost. Composting using the Bokashi method requires lots of water, mixing up organic products, lime and milk sweepings (milk powder) for nitrogen. He found free-range chicken manure, pig manure, sawdust from Stillwater, rotten balage and juicing and clearance apples from Loburn.

Bokashi is a fermentation process, like silage, which reduces wastage. It means that if you are using 100 tonnes of resources, it produces 100 tonnes of compost. Compare that to conventional composting techniques where up to 70% is lost because internal temperatures climb over 60C. He'll turn it over twice, pat it down, cover and leave 12 weeks, then hire a muck spreader to put it on the farm. It's gone from the soil surface within a week.

With Murray, he made 120 tonnes last season. It costs $26/tonne to make and apply on pasture, similar to solid fertiliser, but the benefits are amazing. Pasture roots push down below 30cm and drought-proof the property.

Tim finds composting requires prioritising time over other farming activities, otherwise it never gets done. That said, over time he's developed easier ways to being time-efficient.

When Tim retired five years ago, he leased the 200ha out to Murray Weaver who runs 1,200 sheep including hoggets, plus 150 heifer calves. Murray mates hoggets and gets a 140% scanning result while the ewes reach 199%. Hoggets weigh 45-50kg/lwt and ewes are 80-100kg. Lambs are sold fat with 300 kept for replacements. Murray's focus is on feeding from lamb drop to weaning to get them away early. He wants ewes in good condition by Christmas as that affects next year's weaning.

> That practice transferred all fertility to stock camps and high points, and produced lots of tag where animals hadn't grazed – Max Nichol.

Changing lambing dates made a difference. Tim lambed on September 19 while Murray lambs a month earlier so lambs are gone by Christmas. Pasture is usually pumping by the second week of September. He has very few bearings and some sleepy sickness and must be especially mindful of cast ewes as so many carry multiples. Determining feeding and lambing dates by changing the ram crayon every eight days is much more precise and cheaper than scanning.

Tim reflects on how farming became fun. He observes healthy stock are happy and they'll talk to you. "Now," he says, "the place is looking better than when I finished farming. How many farmers can say that!"

Eco-profit Observations

- Surrounded by high-performance farmers provided tools to step up. If lamb prices had remained constant, profit would have lifted 50%.
- Effective Micro-organisms to replace fish oil. It was more versatile as could spray it on soil, add it to water troughs and drench stock.
- Worm drenches strip gut microflora, so he now drenches EM 5-10ml/hd after as a probiotic to repopulate lamb guts.
- As EM is a soup of living biology, it can survive a couple of months before spreading. It can be applied in the rain so offers tremendous flexibility.
- Bokashi is a fermenting process, like silage, which reduces wastage. Means using 100 tonnes of resources, produces 100 tonnes compost.
- At $26/tonne to make and apply on pasture, similar to solid fertiliser.
- Tim found composting requires prioritising time from other farming activities, otherwise, it never gets done.
- Determining feeding and lambing dates by changing the ram crayon every eight days is much more precise and cheaper than scanning.

9. The Cost of Overgrazing

Greatest contribution to agriculture comes from the county agent or farmer who had the power to observe, imagination to speculate and the logic to deduce a process from which vast benefits have developed.
Louis Bromfield (1948). Out of the Earth.

In early 2025 AgResearch scientists were itching to survey the world's most professional overgrazers - New Zealand farmers - on pasture longevity. But could an increasing butterfat price make such a survey redundant?

Dr David Chapman's Brougham Lecture in 2023 highlighted how production advancements over 40 years largely rested with nitrogen use. But as quoted in Farmers Weekly January 23, 2025, AgResearch's Southland site has shown no sign of pasture productivity gains, Waikato is standing still despite major increases in nitrogen inputs, and Southern Hawke's Bay is going backwards in pasture yield.

While the nitrogen bubble is bursting, Dr Chapman let the cat out of the bag right at the end of the lecture when he mentioned sub-optimal grazing. Could this be where to look regarding pasture longevity?

> There he set up a holistic resource laboratory for students to learn relationships between livestock, cropping and budgeting while challenging traditional thinking – Finding Beneficial Relationships.

Sub-optimal grazing occurs at tiller leaf stage two instead of three because farmers are paranoid about losing pasture quality. It's well established that sub-optimal grazing causes pasture burnout because carbohydrate reserves in plant roots and crowns never fully replenish from severe grazing. This is made worse by plate meters averaging plant height, resulting in shorter pasture recoveries which disadvantages the most nutritious and productive plants livestock graze lower than average height. Using averages reduces pasture and livestock production.

Does anyone ever look at the cost of overgrazing? Grazing coach Siobhan Griffin looked at pricing the effect for dairy farms grazing ryegrass at the two-leaf stage instead of three-leaf. DairyNZ fact sheets say the yield difference is 1.1 tonne/ha over 12 months on irrigated pasture. How could that impact a 250ha property with 25ha in fodder beet?

Assuming average herbage production is 10 tonne/ha, if annual production for two-leaf on 225ha is 2,250 tonnes and three-leaf at 11.1 tonne/ha is 2,775 tonnes, the annual opportunity cost of grazing at two-leaf stage is 525 tonnes. With a ratio of 12kgDM to 1kg milk solids, that equates to 43,750kgMS, or at current prices another $437,500 income.

Siobhan argues farmers plant winter crop because they're overgrazing. Assume 25ha of fodder beet with the paddock out over October and November for successive seasons. Taking the above annual figure, adding an additional 25% for a second spring, comes to a 14-month total production of 2,813 (two-leaf) and 3,469 (three-leaf) tonnes, with a gap of 656 tonnes for fodder beet to fill. That results in a 25ha crop requiring 26 tonne/ha to meet an all-grass system while costing $2,500/ha. Grass is free. If annual forage production is 15,000kgDM, then 25ha fodder beet requires yields closer to 40 tonnes to equate a three-leaf grazing system.

And there's the impact of fertiliser. Ever notice how high nitrogen application rates result in tillers without structural integrity? They kink over an inch or so below the tip which increases the need to graze earlier than three-leaf because quality is dropping. This contributes to pasture burnout and expensive renovation.

All this suggests herbs were the go-to plants when grasses and clovers were below their best – Bland Pastures create Bland Food.

Furthermore, sub-optimal grazing combined with nitrogen fertiliser lifts crude protein levels into the range of 24-28%, sometimes higher. What's never mentioned is how high crude protein overloads haemoglobin with ammonia creating animal health problems and reducing production and fertility. Compare this to North America where authorities impose fines on feedlots shovelling more than 16% crude protein into livestock, even shutting down facilities at 20%.

Remember DairyNZ stating two decades ago how protein was everything and butterfat irrelevant? When Fonterra recently increased its Value Component Ratio from 1.06 to 1.30, meaning the fat price is 130% of that for protein, it triggered an overnight scramble for Jersey AI straws reflecting the breed's higher butterfat production compared to Friesian. As the component composition ratio changes, lifting 60 cents for fat and dropping 78 cents for protein, farmers are rethinking production.

Every farmer knows increasing fibre pushes up rumen acetic acid levels. That drives fat production. Longer rounds generate more fibre and carbohydrate as pasture matures. Ever notice System 5 dairy farms mostly use maize, hay, molasses and grain, all carbohydrate based supplements, the most expensive part of the ration? These costs are why balancing diets becomes uneconomic once crude protein lifts beyond 17%. This is why finishing livestock for slaughter involves hard feed, not lush green grass.

To back this up, recent Irish dairy research is proving diets at 17% and 15% crude protein do not lower milk production. However, what these diets do is lower nitrogen excretion and improve animal health because body fluids remain closer to neutral than alkaline pHs, reducing susceptibility to bacterial problems. This all adds to bottom lines.

Modern research shows recoveries from 5-7 weeks still have crude protein levels over 12%, (remember this is what most finishing feedlots are operating at and most animals cannot process more than 15%). Even a lamb overexposed to crude protein will start slowing its growth rate by day 42 from birth, according to Mark Bader of Free Choice Enterprises, Wisconsin.

American grazing guru Jim Gerrish highlights this too, but looks at grazing too short. Grazing pasture to 5cm height results in 64 days to grow a tonne of feed per hectare. But allowing that residual to grow twice as high reduces the time to reach a tonne to 40 days and over a 200-day growing season would grow five more tonnes than grazing short. Five more tonnes per hectare equates to 500kg of lamb, or at $8.50/kg carcass weight, another $4,250/ha, assuming 100% efficiency.

Will butterfat prices push farmers to experiment with three-leaf grazing? Any dairy farmer grazing “silage grass” with a pre-grazing residual of 3,500kgDM/ha or even higher knows, production doesn’t nose dive, as professionals believe. Maybe AgResearch will study that?

Eco-profit Observations

- Pasture production advancements over 40 years largely rested with nitrogen use but New Zealand research sites are finding production has stalled.
- Eliminating sub-optimal grazing would increase pasture longevity. Pasture meters encourage sub-optimal grazing because the most nutritious severely grazed plants never fully recover.
- Grazing ryegrass at two-leaf stage instead of three-leaf reduces pasture production by 1.1tonne/ha/yr on irrigated pasture.

- Overgrazing necessitates winter crops which lifts costs.
- Nitrogen fertiliser can reduce grass tiller integrity necessitating shorter recovery periods.
- Crude protein in excess of carbohydrate energy ties up haemoglobin and reduces metabolism.
- Current New Zealand butterfat price is 130% of current milk protein price, and farmers are buying genetics to take advantage of that.
- Longer rounds increase pasture fibre, lifting milk fat production.
- Irish research shows dietary protein of 17% and 15% does not reduce production volumes.
- Longer post-grazing residuals grow more grass.
- Pre-grazing residuals of 3,500kgDM/ha don't reduce milk production.

Adaptation: Will Price Hikes alter Grazing Operations? New Zealand Farmer, August 2025, p10.
https://www.pressreader.com/new-zealand/nz-farmer-9bp8/20250729/281621016379543?srsltid=AfmBOopnlveyuxg0XG-7DR9jAWUjj0a-EZ2bgFoPGknPMRnQWhHA_jXf

10. Jeremy Talbot - Balancing Pasture Protein

Semi-retired South Canterbury machinery dealer and contractor Jeremy Talbot has spent a lifetime with livestock production scientists. But he believes good science stopped when Lincoln University's dairy farm changed from best international practice to a pasture-only demonstration farm in 2005.

Initially an A1 mechanic, he ended up farming after diesel poisoning in 1979 and then built a contracting business. That took him into silage production and harvesting, leading to a deep understanding of soil function and its links to crop and livestock nutrition.

He argues what's missing today is Nutrition 101 as few farmers using devices have any idea how things go wrong. For example, no ruminant can utilise a diet of more than 15% protein. Traditionally, Canterbury pasture peaked at 26% protein in October. Now that occurs in February, with Lincoln University recording up to 40% protein in the diets it measures.

Jeremy explains how increasing pasture nitrogen affects nitrate excretion rates. At 18% protein on 18kgDM/day, a cow excretes up to 140 grams of nitrate. At 22-23%, that rises to 500g, at 28% it's a kilogram, over 30% it's 1.5kg with 40% squeezing out 2kg. This is why farmers on Canterbury's irrigated dairy farms no longer see cow pats with a dimple.

> Longer rounds result in cows consuming a higher fibre diet which reduces many animal health problems – Karl

Driving this outcome is ryegrass succumbing to heat stress at an air temperature of 25C, but under irrigation what does grow is clover, literally pure protein. This is why cows are now emitting 190% more nitrates into soil than they did 30 years ago and suffering with metabolic problems and fertility issues.

Industry advice is to substitute with DDG, sunflower seeds and other high-protein feeds which poison cows further. Farmers using PKE see a slight improvement because at 18% protein its dialling down pasture protein of 25%. PKE can be as low 8 ME, especially if treated with hexane to extract oils. Excess PKE increases copper toxicity and reduces butter integrity as it melts at room temperature. Is this why DairyNZ limits PKE to 3kg/hd/day?

Excess dietary protein costs the dairy and meat industries on many fronts. Jeremy found Fonterra buys up to 100,000 tonnes of lactose annually from

American corn-fed cows to mix with New Zealand milk powder to meet international standards. The lack of pasture carbohydrates is why Atkins Ranch says New Zealand farmers will never achieve meat premiums until they feed livestock better. Finishing livestock on excess protein, similar to an Atkins diet, strips meat of fats. Animals never marble.

The trend in Ireland and Europe is to restrict diets to 17% protein, with Irish research backing a drop to 14% and increasing milk production. Any C4 grass is low-protein and modern maize varieties can grow 25 tonnes of dry matter versus 5-6 tonnes of pasture in 3-4 months, providing low-protein feed by only using stored soil nitrogen.

Here's a simple formula to solve methane and nitrates without vaccines, GMO, seaweed or a stack of technologies promoted by rural media. Lincoln University professor Pablo Gregorini shows balancing dairy diets with carbohydrates can knock back nitrates up to 80% and methane by 30%. At the same time, production went up 25%. Is there a single farming advocacy agency listening?

Back on the farm in the early 1980s, when faced with spring protein excesses, Jeremy started harvesting cereal silage using ideas from the UK. The family farm was 600ha, so he experimented there with gear and techniques. In building his own cultivation equipment he swears nothing beats hours in a tractor cab observing the results of what he designed.

> This pushed red clover everywhere. It spread on its own and grew no matter the climate. Its tap root broke up any pugging and kept its quality all the way into winter – Guy and Ashley Coats.

Soils expand when wet and shrink when dry. This is how micro-cavities develop and help nutrients get to plant roots. Irrigation leads to compaction, but agencies associated with land management and science ignore its root cause.

Compaction holds all dung, urine and fertiliser under soil surfaces in layers, creating stratification, meaning plants grow in a urea pan. Ploughing breaks up stratification and mixes layers of nutrients within soil profiles. AgResearch studies show turning soil over sequesters carbon stuck at soil surfaces. Globally, more farmers are using strategic ploughing to improve soil structure and to balance nutrients throughout soil profiles.

Everybody knows you can't make good silage from pasture fertilised within 4-6 weeks, but a soil urea pan produces the same outcome. When crude protein is high, it becomes ammonia in silage stacks, killing silage-making bugs. This process happens in livestock rumens with ammonia clogging up blood haemoglobin, leading to metabolic problems. Lower protein levels significantly improve livestock health when pasture recoveries extend from 21 to 35 days. Coincidentally, the international standard for silage making is 5% seedhead.

Jeremy joined with UK professor Mike Wilkinson and Lincoln and Ruakura academics, as well as several dairy consultants in the Better Sustainable Farming Group. They refer to themselves as the Flat Forehead Society after struggles with DairyNZ and others to acknowledge damage from industry advice to farmers' incomes and the environment. Jeremy argues there are major credibility problems with New Zealand research institutions because of how they are funded and led.

For example, trials using plantain to reduce nitrate emissions were cancelled after three weeks because cows wouldn't eat it and lost condition. Yet a paper emerged saying 5% in pastures addressed nitrates in livestock production. That finding was incorporated into Overseer, the nutrient budgeting programme mandated for all farmers. A recent $ 22 million study confirms plantain doesn't reduce excreted nitrates.

It's this research that has him challenging national science policy. The drive for all research initiatives to be income positive and to tax farmers for problems they don't have is the final nail in public-good science.

Eco-profit Observations

- No ruminant can utilise a diet of more than 15% protein for very long.
- Increasing pasture nitrogen affects nitrate excretion rates.
- Dairy cows are now emitting 190% more nitrates into soil than 30 years ago.
- Excess PKE increases copper toxicity and reduces butter integrity as it melts at room temperature.
- Fonterra buys 100,000 tonnes of lactose from American corn-fed cows to mix with New Zealand milk powder to meet international standards.
- Ireland and Europe is restricting diets to 17% protein, with Irish research backing a drop to 14% and increasing milk production.

- Professor Pablo Gregorini shows balancing dairy diets with carbohydrates knocked back nitrates 80% and methane by 30%. Production lifted 25%.
- Compaction holds all dung, urine and fertiliser under soil surfaces in layers creating stratification, meaning plants grow in a urea pan.
- Strategic ploughing improves soil structure and balances nutrients throughout soil profiles.
- Can't make good silage from pasture fertilised within 4-6 weeks, but a soil urea pan produces the same outcome.
- International standard for silage making is 5% seedhead.
- Trials using plantain to reduce nitrate emissions were cancelled after three weeks because cows wouldn't eat it and lost condition.
- Major credibility problems with NZ research institutions because of how they are funded and led.

11. Henry Hendricks - More from Less

Imagine dropping fertiliser use by 70% only to see production increase by 10%. That surprise greeted Henry Hendricks, a dairy farmer on the outskirts of Hamilton.

Henry is proudly a System 5 farmer, milking 280 cows on 64ha producing over 140,000kgMS on a lactation of 300 days. He harvests a minimum of 16,000kgDM/ha annually from pastures to add to 940 tonnes of bought-in supplements. The cost of fully feeding milking cows with supplements is $3.85/hd/day, around 6kg/hd/d. Supplying 10kgDM or more daily doesn't make money.

> During her veterinary training she had little exposure to nutrition, let alone any appreciation of what nutrition can do for improving livestock health – Louise Coats.

You might think animal health would be an issue with these production figures but the herd only sees vets for a lepto injection and applying dry cow therapy. He had four empty cows after a nine-week mating last spring (2023) which coincides with milk urea nitrogen (MUN) in single figures. In fact, MUNs barely reach 20 throughout the season.

Mating is over nine weeks using sexed semen. It takes 15 days to use up all the straws, then the cows go out to the bull for natural mating. He inseminates from 4.30pm in the afternoon after milking. He will keep a bull calf from his herd for natural mating and use it for one season.

He achieves 80-95% calving within three weeks, delivering 17-20 calves a day, which means he needs a lot of feed on hand. This is when cows come out of the winter barn and overnight on pasture. He only needs 40 heifers for replacements (14%). Any cow that annoys Henry is culled, whether it be temperament, succumbing to mastitis twice or reaching older than 20 years.

This indicates cow health is excellent because they're not over-fed protein using software like Farmax. He doesn't need electronic collars as he is always out with the mob observing them. Cows are the boss. They always have more grass and quality feed, meaning Henry's urban neighbours aren't plagued by smelly dung. Technologies often reduce farmers' abilities to work with nature.

So, what else is surprising? Grazing recovery doesn't drop below 32 days. Why? Cows prefer a slower round and eat herbs (weeds). In fact, the first round is 70 days; it doesn't finish till the first week of October when many dairy farmers are easily on their third. If wet after drying-off, cows are fed supplements to stay off paddocks and grow high pre-grazing residuals for first round.

Otherwise, cows are fed 14-15kgDM grass plus 6 kg of supplement during lactation. For this to happen cows have pre-grazing covers of 4,500kgDM/ha and leave 1,500-1,600 so that difference determines round length. They're never treated for bloat.

Could high pre-grazing covers be part of his secret? Henry would be one of the few dairy farmers who grazes consistently at the third leaf stage, maybe higher. Farmers and professionals overlook this all the time. He observes most dairy farmers panic when pastures reach 2,400-2,500kgDM and skip them for silage. They harvest 10 days later for perfect silage whereas he would rather feed that to his cows. He asks, "How can farmers grow 12.3ME silage but not fully feed their cows?"

Furthermore, there's a savvy business reason. He argues that by grazing "silage grass" he can milk an additional cow and a half every day for no cost. That comes from experience, not feed-budgeting software. Higher covers also reduce irrigation requirements.

As a result, peak milk occurs in November and December, much later than other dairies. What does he put this down to? A long first round and by its end the topper comes out to drop grass in front of cows which coincides with mating. Every paddock cows get from then on has a rising nutrition plane. At Christmas they're pumping 2.2kgMS/day. His later peak is longer so he produces more milk solids.

> With production lifting 30% in three seasons by sowing deeper rooting pastures, enhancing soil fertility, longer summer rounds and focusing on productive heifers, they demonstrate getting basics right is key to developing a successful, simple operation – Peter and Christine Johnstone.

So, is pasture renewal part of his success? No. Take a sneak peek at his paddocks and you'll count every weed under the sun. Henry believes in working with what you've got. Buying feed is lower risk than ripping up pastures. Why? Because it takes five years for paddocks to recover from being over-stimulated growing maize.

When it rains, pasture recovery pushes out to 40-50 days with balage, maize and PKE. This also saves peat soils from over-trampling. High fertility soils pug easier, especially as earthworm numbers rise. Within 14 days there's grass everywhere and the pre-grazing mass climbs above 5,500kgDM. Cows eat their plate clean.

All maize is kept at the shed. He dropped 100 tonnes this year and replaced that with lucerne balage. The herd gets a bale a day, costing $170 each. Protein is usually cheaper to buy than energy. The idea came from North America on social media. Americans use maize, canola and lucerne these days. Henry finds shifting to something different stimulates his mind and business, otherwise, as he says, it's the same shit, different day.

So, what about this fertiliser decrease? His secret is simply changing from solids to solubilising urea, sulphate of ammonia and potassium chloride and spraying. Some 4,000 litres of water with 500kg fertiliser plus a bit of Maxicrop covers 18ha. Instead of one or two dressings per year, he does 20 and has dropped from 220 units of N to 80. Including a mixing system, his fertiliser bill fell $100,000 in 2022, from applying 12 tonnes per month to three. Pasture production dropped from 18,000 to 16,000kgDM but cow production rose from 500 to 550kgMS/hd.

Will Henry ever be recognised for his discoveries? Unlikely. No fertiliser company ever promotes clients using 70% less product. No bank promotes clients with farm working expenses below 50 cents in the dollar. DairyNZ wouldn't promote anyone with pasture less than 50% ryegrass. No regional council would promote a farmer running 4.7cows/ha.

Eco-profit Observations

- Supplying 10kgDM supplements or more daily doesn't make money.
- Low empty rates coincides with single-figure MUNs.
- Inseminates from 4.30pm in the afternoon after milking.
- Achieves 80-95% calving within three weeks and needs a lot of feed on hand.
- Only requires 14% replacements.
- Grazing recovery doesn't drop below 32 days, cows prefer a slower round and eat herbs.
- Pre-grazing covers of 4,500kgDM/ha and leave 1,500-1,600 so that difference determines round length.

- Grazing "silage grass" results in milking an additional cow and a half every day for no cost.
- Peak milk occurs in November and December due to a long first round resulting in a longer peak so more milk solids.
- Purchasing feed is lower risk than ripping up pastures.
- With rain, pasture recovery pushes out to 40-50 days with balage, maize and PKE to prevent peat soils pugging, which increases as earthworm numbers rise.
- Simply changing from solids to solubilising urea, sulphate of ammonia and potassium chloride and spraying dropped from 220 units of N/ha to 80. Fertiliser bill fell $100,000.
- Henry will never be recognised for profitable interventions.

12. Cliff Shearer - Passion for Production

As an historian of agriculture, Cliff Shearer will tell you there are no new ideas. The first rotary cowshed was American, milking 1,700 cows in 1921; chicory was used in UK pastures in the 1880s; English farmers used PKE in the 1880s as cake oil. Born an organic farmer, he bought all the organic books and has an extensive library of agricultural classics, yet he lives for production.

Cliff's passion for achieving incredible milk production with a Jersey herd of up to 30 cows on an 8ha milking platform is well known beyond Hawera, Taranaki. His cows regularly squeeze out more than 1.5 times their liveweight in milk solids (730kgMS/cow in the 2022/23 season). His interest is getting the herd average up, not in cup winners or high-producing animals.

All cows milk for 305 days, then get 60 days off. His highest-producing cow recently squeezed out 1,062kgMS and his highest heifer 800kg.

Production is an addiction. He's herd tested every month for 40 years! This is a serious System 5 dairy farm and a classic example of how fully-feeding livestock achieves extraordinary production. It also highlights how such systems are exposed to inflation and risks. So, what drives this production?

Feeding is the primary driver but Cliff cautions, feed cows to match your breeding. Cliff believes in fully feeding his animals. They're supplemented 16kg, half in the shed and the rest in the paddock. It's a grain mix from New Plymouth, essentially a carbohydrate muesli containing soya hulls, DDG, PKE, molasses and starch pellets. Clearly, production is carbohydrate-driven.

His genetics focus is optimising feed conversion. He states, "If only on silage and grass, my cows would never reach 450kgMS, maybe 400 at the most." If he did go all-grass, the reversion would take two years of fully feeding to return to the original production of System 4 or 5. This insight highlights how long it takes for any herd to really hit their straps when changing feed. He says, "You've got to breed for your feed and then feed your breed." Jerseys being smaller cows are more efficient because they use less feed for maintenance.

Within two years of stitching in plantain, stock no longer needed cropper drenches saving time and money – Peter Coles.

But this system comes at a cost. Feed expenses doubled

over two years (prior to 2022) essentially gobbling up two-thirds of his herd's income. Also, relying on importing supplements has left him vulnerable to metabolic issues when mixed wrongly. Excess ruminisin, copper, zinc and magnesium oxide nearly exterminated his herd at least twice. What really hurts is that often the best cows eat more in trying to self-medicate only to speed up their own demise. Blending feed is about the only activity he doesn't do himself and reflects farm scale.

In 1979 he started a Jersey stud and the history books soon became invaluable. His favourite farming hero is Englishman Robert Bakewell, whose use of line-breeding and inbreeding doubled animal size in the 1770s. As Cliff's herd largely derives from one heifer bought in 1980, he's a great believer in line-breeding. While many dairy farmers shake their heads at a lack of BW values, they can't ignore cow production. He has a strong following on Facebook and sells straws to many clients who find him there.

In studying the old techniques of Bakewell and others, Cliff argues high-producing heifers become high-producing cows due to their pedigree, whereas New Zealand BW values are per-hectare based so are irrelevant for System 5 farms. When comparing LIC genetics to his own hand-picked bulls, cows from his bulls produced 60kgMS more.

Bull selection starts by picking mothers from outstanding calves. "They just seem better than the rest." He says he relies on intuition. "Next, get her pregnant and keep the bull." He's used over 500 names starting with C for his bull line, and is now starting to reuse some. Cliff believes farmers are best to do their own AI but it's expensive to hold straws as they cost $8 and another 50 cents to store them. Cliff has 14,000 in storage.

Cows producing over 800kgMS determines which bulls he keeps. If he was ever to buy a bull its dam would need to do over 750kgMS regardless of BW because that's his herd's average.

Herd turnover is high with half the herd culled to accommodate heifers. Any opportunity to cull and they go. It's a constant activity with spring and autumn calving. All heifers are grazed off-farm, otherwise he'd be milking 15 cows.

What makes this achievement remarkable is the lack of fertiliser and agrichemicals and little pasture renovation in 40 years. During his first year farming, a salesman said he had to use fertiliser. He did half the farm but didn't see a difference and he has been waiting in vain ever since. He states, "The only place to experiment is your own farm."

Animal health is simple for Cliff - "feed more hay," an idea from Bill Paterson, a Pennsylvanian grass-based dairy farmer. Animals get hay every day and more if under the weather.

Everybody coming up the drive had a better car and knew more about their business than they did. How do we get to that? he asked. How can we turn it around? – Tim Chamberlain

Cliff's journey certainly demonstrates that focusing on genetics and feeding can deliver amazing production. His experiences also highlight the vulnerabilities of being small and importing feed. While selling semen and cull animals provides secondary income, cash surpluses from milking are compromised by scale and feed cost inflation. He is the only small-scale dairy farm along his road whereas 12 months earlier there were another three.

There's no doubting Cliff's passion. Upon contacting Cliff for this story, he'd milked twice daily for over 5,500 days straight. That must be a national record.

Eco-profit Observations

- No new ideas, plenty of history repeats itself.
- Milking platform eight hectares, herd is 30 cows.
- Average animal production target is 730kgMS/hd, one heifer has reached 800, highest cow 1,062.
- Lactation is 305 days; all cows get 60 days off.
- If dropped all-supplements for all-grass and then switched back, it would take two years to return to the previous all-supplement production.
- Jersey cows have a smaller frame, therefore less maintenance, therefore more feed-efficient.
- Supplying two-thirds of feed through supplements opens up the risk to inflation pressures.
- Blended supplements are fraught with risks, especially herd poisoning.
- His herd lacks BW values, but is sold on production volumes. His hand-picked bulls daughters produce 60kgsMS more than top LIC bulls. They come from cows producing 800kgMS or more.
- Turnover half the herd each year, all heifers grazed off-farm.
- No fertiliser, agrichemicals or pasture renovation in 40 years.
- Animal health remedy is hay.

13. Phil and Viv Gray – Grazing Increases Flexibility

Modern production relies on a continuous water supply, but what happens when the tap is turned off? How Phil and Viv Gray dealt with this mirrors the experiences of dryland farmers looking for another path than monoculture lucerne.

Phil started farming with his father Hugh in 1983, running 2,100 ewes and replacements plus 100 trading cattle, at their property, Longford, near Waiau, North Canterbury, an area prone to summer dry and drought. Problems started when they played to the farming industry's desires rather than sticking to their own game plan.

Phil recalls using grazing practices similar to Holistic Management but didn't understand what worked and why. He had no plan of what he was trying to achieve with his business when putting all animals into one mob or lengthening recovery times between grazing rotations. As a result, Phil intensified animal production and set-stocking.

"I began focusing on lamb performance to the detriment of everything else. As a result, the ewes packed up and we hit the wall." Lambing performance dropped from 148% to 110% in two years. He admits overgrazing too many paddocks and spending way too much on superphosphate, creating a copper deficiency that caused ewes to lose condition. That took three years to correct. They refocused their stock policy to prioritise ewes and worked around them.

At the same time, they regrassed the farm to deal to severely debilitating grass staggers. For seven years from the early 1990s they ploughed and double-cropped half the farm at great cost to replace ryegrass with fescues. Yet within 18 months of sowing fescue, ryegrass was back.

"It was very demoralising," Phil recalls. "We were spending our way out of debt but digging ourselves in deeper. Everywhere I looked I could see the ryegrass coming back and there seemed no way of getting on top of it." Looking back, he realises there was probably over a hundred

> That got him renovating pastures without herbicide by severely grazing first and then direct-drilling chicory, plantain, radishes and white clover – Dennis Nuiewkoop.

kilograms of high-endophyte ryegrass seed per hectare. "What were a few kilograms of fescue seed ever going to achieve?"

During this time they used cocksfoot, chicory, timothy, plantain and prairie grass in pasture mixes. "Species like timothy and prairie grass never came through. We only started to see their presence across whole paddocks after we changed to holistic planned grazing." After realising eradicating perennial ryegrass by cultivation was impossible, they considered it would be less risky to use grazing management to create multi-species pastures.

Phil noticed ryegrass domination reduced when they changed to longer pasture recoveries to favour deeper-rooting grasses such as prairie grass and timothy. Red clover appeared in paddocks it'd not been sown in for 20 years. Greater pasture diversity lengthened their grazing season and improved animal health and performance. Ryegrass staggers disappeared after changing grazing management.

"We started thinking breed has far less to do with overall performance of the farm," says Phil. He recalls the saying an animal is 10% breeding, 90% feeding. "If we get our soils right, balance stock protein/carbohydrate and mineral intake and have year-round access to good quality water, our grazing management and willingness to drop lambs early gives greater flexibility to see things through. If we have grass to finish lambs that's great, but if not, our ewes must come first."

Selling store lambs dropped animal health bills and reduced pasture contamination. As a result, they changed their grazing policy to adapt to the conditions. "The cheapest way to grow grass is to mob up," says Phil. He describes how he employed this policy in a cold dry spring in 2002. "In mid/late October we put all the ewes, lambs and our bull beef in one mob and began with a recovery period of 60 days around the farm. With the predicted dry summer and slow spring growth, mobbing up was the only way we could build a grass bank ahead of us."

Everybody else in the district remained set-stocked. "Moving away from set-stocking helped us grow more grass cheaply and we had less stress finding money to pay fertiliser bills," says Phil. "We sold half our lambs as stores in November at $5 above that budgeted and dropped R2 bulls at Christmas as normal. We bought bull calves then, thinking they would handle the dry better." About then they lengthened their recovery period to 110 days. Looking back, they feel they got their stock policy right for the spring and early summer.

Landscape planning improves farm efficiencies by identifying locations of fences, shade, shelter and clean water – Bill and Sue Garland.

"We sold most of the lambs in January 2003 and then came unstuck in February when our creeks dried up," he says. They should have sold the rest of their lambs then as they struggled to get water up to the mob and growth tanked on hard feed. Behavioural problems between cattle and sheep together meant days were busy around water troughs. By repeatedly returning to many smaller mobs when creeks ran dry, overgrazing of the better plant species saw pasture production decline over several years.

It finally rained late March and within four weeks pastures averaged covers of 2500kgDM/ha across the farm. "We never had so much feed coming out of a drought," Phil said. They bought calves in May and sold lambs at prices triple that offered in February.

Recalling those days, Phil suggests dryland farmers invest in water systems to cope with large mobs in dry times (use fertiliser budget to fund that) and target sheep genetics such as Ultimates and Texels that grow on mature feed. And instead of maintaining pasture quality across the whole farm, use deferred grazing to keep good quality in smaller areas to optimise lamb growth and defer different paddocks each season. Such practices increase soil carbon and biology to lift soil nitrogen stores and prolong pasture growth with available moisture.

Eco-profit Observations

- Singular focus on lamb growth rates and finishing crashed ewe health and performance.
- Too much superphosphate created copper deficiencies in livestock.
- Similarly, focusing on removing ryegrass to address ryegrass staggers proved ineffective because the ryegrass seed bank outlasted any other species sown.
- Only by lengthening grazing recoveries did later-flowering species such as timothy and prairie grass survive.
- Changing grazing recoveries allowed red clover to emerge in paddocks despite not being sown for 20 years.
- Ryegrass staggers disappeared after changing grazing management.
- Focused less on breed and more on feeding of capital stock.
- Moving from set-stocking to mobs grew more grass cheaply, used less fertiliser and pushed a wave of feed into summer.

- Stock water quickly became the limiting factor when running one large mob during the dry. That led to smaller mobs, more overgrazing and eventually reductions in pasture performance.
- Grazing the one mob produced significant pasture production once drought broke.
- Prioritise stock water systems first, have hardier animals that produce in the dry, use deferred grazing to maintain pasture quality for lambs and use grazing management to lift water and mineral holding capacity by increasing soil carbon.

Adaptation: Blue Skies for the Gray Family. Mar/Apr 2005. In Practice (100) p.11. Holistic Management International.

14. Will Perriam-Lampp - Focus on Utilisation

Utilisation, not yield, is where money is made feeding livestock. Increasing the amount of grown feed going down animals' throats is better for business. However, many sales reps offer advice based on what they've been instructed to say, rather than actually finding out what farmers want to achieve. They often know more about sales than farm systems.

Journeyman Will Perriam-Lampp has worn many hats in the service sector after a rugby injury made him look for a career beyond dairy farming. He advocates farmers having robust conversations with reps about their needs. Here are some utilisation insights from his current role as manager of Align Farms' dry stock farm near Methven.

> If it's worth $500 and sitting under a tree, it's out the gate and the cash is put into more productive investments – Simon Lee.

Preparation is everything. If direct drilling, best to deliver soil surfaces ideal for either tyne or disk drills to be used. Tyne drills don't handle crop trash. Will's seen 16 tonne/ha swedes sown from helicopters, so cultivation isn't always essential. In New Zealand, only 35% of farmers plough with two cultivation passes and 75% of that occurs on the Canterbury Plains. Old habits die hard.

Plant germination is driven by seed-soil contact. A Whanganui farmer demonstrated this grazing lambs on 100ha of rape. Four days before pulling the lambs out he helicopter broadcast Italian ryegrass and urea for lambs to hoof and tooth. That combination stimulated a good strike, creating lambing pasture for ewes.

Being sold on cultivar yield is a trap for farmers. Any fodder beet crop needs its utilisation considered; no point having the highest yielding cultivar if 50% remains underground. If livestock access 60% or more of bulbs from lower-yielding varieties, it'll lift utilisation and returns.

Farmers can check this themselves while checking plant populations. Will visited a Roxburgh farmer whose very large swedes achieved lower hectare yields than a smaller bulb variety with higher numbers per square metre - 14.5t/ha versus 16. Wastage from heavy woody stems and cutting fence lines in kale crops is often ignored when quoting expected total tonnages.

For pasture utilisation, think flowering and heading dates that affect relative maturity when planning pasture mixes. Reconsider high-yielding ryegrasses for those that fit a site's growth curve better. Later heading dates of higher-yielding cultivars create feed pinches early spring. Fescues have autumn dormancy, so are not good for pushing feed forward for winter. With a summer or winter fodder crop, does a mix need a 60 or 120-day turnip to suit planned grazing dates?

With pest issues, farmers need to calculate if sprays cost more than insects eat. Spraying might lift costs from 7 cents/kgDM to 14 cents (when yielding 6 tonne/ha) but it's still cheaper than buying feed at 34 cents. With caterpillars, 2-3 per plant is enough to be concerned. Check for diamondback moth on underside leaves, white butterfly on top and aphids in growing points. Look for anything parasitised (brown body with a hole) as parasitic wasps are lurking. Sprays kill them too.

He had observed soil degrading under urea and pastures always needing more nitrogen. He started by dropping superphosphate for RPR – John Vosper.

Stressed plants emit infra-red signals to attract insects, so if you are considering spraying walk well beyond gateways as compaction lifts plant stress. Will observed a raphno crop planted in a paddock of sandy and silt soils. Diamondback moth decimated the plants in sandy soil due to moisture stress while the plants growing in silt were perfect. This means a rain event reducing crop moisture stress and lifting plant sugar levels is as effective as a pesticide application. Why rush for a spray can when rain radar shows a storm brewing.

Catch crops provide opportunities for pastoral farmers to consider. Cereal cropping systems sow autumn catch crops to suck up nitrogen before terminating in spring for another crop. The opposite is a legume crop. Will calculates sowing winter faba beans costing \$260/ha could release 120kg of nitrogen upon termination at 4,000kgDM/ha. If nitrogen released equates to \$240/ha, it means bean seeds cost only \$20/ha.

Sometimes, best practice is increasing crop turnover instead of harvesting at full-term. Being dryland, silaging green catch crops in early November maintains feed supply better than maximising whole chop yields in January. This allows for a summer grazing crop to fill a mid-late summer gap or for establishing winter crops while moisture is available.

If ryegrass struggles to strike well after fodder beet, reconsider cropping rotations. Fodder beet to kale to pasture minimises chemical residuals affecting subsequent pasture species like clover (plantback), maximises nutrient recycling and often produces higher yields with cheaper costs. Adding 30kg/ha of oats with new pasture improves soil structure while lowering weed pressure and providing more feed. Oats, being annual, have aggressive roots that penetrate clods.

Utilisation is also important for animal health and drench resistance. A technician pointed out to Will that oral is best. Drench marketing strategies focus on convenience with pour-on easiest but least efficient; animals need to lick each other for it to work. Farmers are never told the efficacies of skin-penetrating ingredients. Such insights dropped his R2 drench bill from $4,000 to $700.

Will's changing his grazing management to address drench resistance. With mobs going onto 3,000kgDM and leaving 1,500kg, he allows R1s to pick the first 500kg, then R2s graze down the next 1000kg as they have better immunity. It requires a bit of juggling but consider the following thought: Farm system specialists calculate that the optimum financial performance from livestock occurs when they are fed at 80%. This allows you to carry 12% more animals with spare feed. Will ponders if drench resistance will challenge that recipe, especially as animals get larger. Organic farmers prove that when animals are fully fed and not stressed, they're more resilient.

Imagination, not money, solves problems. Utilisation cranks up revenues and profit, something Will is judged on when compared to Align's dairy properties. However, he acknowledges fodder crops cause many more animal health issues than all-grass wintering and luckily for him those costs fall on dairy farm gross margins.

Eco-profit Observations

- Utilisation, not yield, is where money is made with feeding livestock.
- Will's seen 16 tonne/ha swedes sown from helicopters so cultivation isn't always essential.
- Plant germination is driven by seed/soil contact. Helicopter broadcasting Italian ryegrass and urea for lambs to hoof and tooth while grazing a kale crop gave an excellent strike rate.
- Check crop tonnages. Smaller and more numerous swedes outperform larger fewer bulb crops.
- Later heading dates of higher-yielding cultivars create feed pinches early spring.

- Farmers need to calculate if sprays cost more than insects eat.
- If considering spraying, walk well beyond gateways as compaction lifts plant stress.
- Rainfall events are as effective as pesticide applications by reducing crop moisture stress and lifting plant sugar levels.
- If catch crops hold nutrients through winter, could legume crops produce enough free nitrogen to reduce fertiliser for the next crop?
- Sometimes, best practice is increasing crop turnover instead of harvesting at full-term.
- Sowing oats with new pasture improves soil structure while lowering weed pressure and providing more feed.
- Pour-on parasite drenches are easiest to use but least efficient; animals need to lick each other for it to work.
- Change grazing to address drench resistance, put young stock in first to graze tops and then mature animals to take second and third bites.
- Organic farmers prove when animals are fully fed and not stressed, they're more resilient.
- Fodder crops cause many more animal health issues than all-grass wintering and luckily those costs fall on dairy farm gross margins.

15. Leveraging Time Lifts Profits

The only problem is that high-leverage changes are usually highly nonobvious to most participants in the system. They are not "close in time and space" to obvious problem symptoms. This is what makes life interesting.
Peter Senge. (1995). The Fifth Discipline.

As farmers belts tighten for a second year, many will be considering which expenses actually make them money. One of the biggest problems for farmers being self-employed is that productivity is limited by their own time and energy.

As Allan Nation, editor of Stockman Grass Farmer, observes about treadmills, you can run harder and make more money, but do anything less and income declines. Enterprises that cannot be time-leveraged will reach their income ceiling faster. So how might livestock farmers leverage their time on-farm?

Reducing mobs is simplest, something dairy does well. For sheep and beef farmers reducing mobs does a number of things. Firstly, it reduces electric fencing, travel and potential time looking for stock, water and electricity problems. It also increases stock density that reduces nutrient transfer from the middle of paddocks to the edges. With larger mobs, paddocks can stay large, reducing subdivision costs. It also makes culling easier as environmental pressures weed out stock not worth breeding from. And it grows more grass.

So how could farmers leverage time further? In 1970s, Southland farmers Joe Guise and Russell Kean put their ewe flocks together for 90 days over winter. This created higher stock density without buying more animals or more subdivision. The mob spent 45 days on each property meaning the other farmer had 45 days off. Seriously, imagine what could a farmer do with a month and a half of Sundays?

In fact, take that a step further. Imagine neighbouring farmers forming a livestock grazing company while retaining land ownership. Profits would be divvied up based on how much time the mob was at each property. Again, farmers only supplied labour when animals were on their land, leveraging free time for neighbouring business partners.

Experiences in Australia, such as Tilbuster Commons, demonstrate multiple benefits for farming families that did this. Included were higher profits and lower costs through economies of scale, but there was more.

Because of working together, families became more proactive with their decision-making, meaning they acted sooner in times of drought. Farmers also looked beyond their own farm boundary when it came to investment as they focused on their company's progress rather than their own self-interests. For example, stock water.

Researchers concluded it's the deep desire to work together and not let each other down that makes farming families proactive under these circumstances. Families become less concerned about issues such as trespass while engaging in conservation or other business projects. Leveraging time strengthens community and profits through developing leadership skills.

But even for individual farmers there are surprising mental health benefits from how grazing management leverages time. According to the 2010 Australian Communities and Landscapes project, replacing high-cost inputs with grazing management was viewed by many as an opportunity to step off the "productivity treadmill" and take greater control of their future. The project interviewed farming families and found certain practices allowed them to change when "locked in" generational traditions were no longer creating positive outcomes. This included dropping fodder crops and regrassing.

> As banks cracked down on seasonal finance, farmers dropped fertiliser applications and started milking their land's natural capital. Many realised they'd been putting too much on – Equity Crisis Lessons.

This fits with observations by ex-Telford farm manager Allan Gorton, that farmers fail to achieve spring production when they stop focusing on feeding (grazing management) and take out more paddocks for new grass and swedes in an already tight system. What does that mean?

Grazing coach Siobhan Griffin uses DairyNZ research to demonstrate this. Sub-optimal grazing at the two-leaf stage instead of third-leaf reduces pasture production by 1.1 tonne/ha on irrigated land. This practice necessitates crops. Yet Griffin's calculations show optimal grazing management at three-leaf stage outcompetes 20-tonne fodder beet crops in tonnage over 14 months without any pests, costs and consents. (Remember Chapter 9?)

Furthermore, some farmers observe crop paddocks take longer to recover, sometimes years, from heavy feeders such as fodder beet or maize. That's not something you'll see in a gross margin. Such crops collapse soil structure resulting in a wall of golf balls in front of power harrows, leading to bigger machinery to pulverise clods.

As UK farmer Jeremy Clarkson demonstrates, expensive machines drive up depreciation but when income drops, that drives down profit. It's about as sensible as pulling the choke out in top gear.

> When Angus and Hereford steers arrive at the same liveweight and go through to the same carcass weight, Herefords arrive at that a month later but ate less feed to get there. Who's looking at feed conversion efficiencies and meat quality? Branding Meat Quality

One big benefit reducing winter crop is availability of spring pasture. Many dairy farmers experience problems with their second round, particularly if cloudy weather reduces pasture growth. Those switching to bale grazing benefit from more spring grass paddocks with similar costs to crops.

Any farmer giving up spring cultivation finds they have time to push a wave of feed forward into summer. Because they're on top of their grazing management, they foresee when feed pinches are coming and destock before markets decline.

Leveraging time is no different to leveraging capital. It allows farmers to progress faster than their own resources. Fewer mobs grows more grass and reduces fertility transfer because of livestock confinement. It's this secret that generates a higher rate of return.

Eco-profit Observations

- Enterprises that cannot be time leveraged will reach their income ceiling faster.
- Fewer mob numbers reduces electric fencing, travel and potential time looking for stock, water and electricity problems, nutrient transfer, subdivision costs, makes culling easier and grows more grass.
- Combining flocks or herds with neighbours is cheap way to leverage time, creates economies of scale, generates proactive decision-making and develops leadership skills.

- Research shows replacing high-cost inputs with grazing management is an opportunity to step off the "productivity treadmill" for farmers to take greater control of their future. This has mental health benefits.
- Heavy winter feed crops result in paddocks taking longer to recover productivity due to collapsed soil structure requiring bigger machinery to pulverise clods.
- Reducing winter crop promotes more spring pasture, more time to plan grazing, greater chance of foreseeing feed pinches.

Adaptation: Beat Inflation with Livestock Management. Farmers Weekly May 8, 2023, p.17
https://www.farmersweekly.co.nz/opinion/beat-inflation-with-livestock-management/

16. Travis Leslie - Change by Numbers

When you know how hard it is to change yourself you can appreciate how hard it is to change others, especially within a corporate. However, corporates can also offer great support, as Travis Leslie discovered when moving to Pāmu's Kepler Farm at Manapouri in 2013.

When he arrived, the traditional 1,650ha sheep-dominated Southland farm had 7,000 commercial ewes, a sheep stud, and 200 breeding cows. As a breeding unit it made money from selling store lambs and cattle to other Pāmu properties. Rainfall was unreliable, despite an annual average of 1,200mm, and this, combined with dense glacial soils, produced results similar to very dry properties.

With management support, Travis challenged the status quo by transferring ideas from up north in the King Country where he had previously managed a Pāmu property. His first test came in the form of drought in his first season when he unexpectedly produced the same number of lambs from 500 fewer ewes. At the same time, he introduced tall fescue and the increased grass production fuelled a shift in management focus towards breeding-finishing.

Historically, only 500 weaned lambs were prime for processing. That year they sent 5,000 lambs to finishing properties and retained replacement ewe lambs. Today, with a 50:50 sheep:cattle ratio and all ewes mated to terminal rams, 70% of lambs are processed immediately averaging 18kg cwt at 100 days weaning and rest sold as stores. In 2024, 2,500 lambs averaged 18.5kg cwt while 1,000 went store at 31kg liveweight. With all lambs gone at weaning, focus can shift to finishing cattle.

A range of benefits emerged from questioning winter crop and overseeing a gradual reduction from 140ha to no cropping by 2020. This wasn't planned. Due to difficulties wintering ram lambs on crop, trialling all-grass wintering proved so successful Travis included two-tooths the following season. Both classes thrived and it didn't lower the stocking rate. Cattle followed in successive years and mixed-aged ewes responded as well, if not even better, than other livestock classes. This was a pleasant surprise.

Other benefits from dropping crop include ewes with better teeth lasting a year longer, less mud, less weekend labour and every paddock available in spring, all creating flexibility. Spring surpluses are made into hay and balage for winter. Even with increases in supplement use, costs reduced significantly. Since 2019, this change means savings of $100,000 annually.

Furthermore, reducing winter crop and changing stock policies lifted Kepler Farm's performance and stocking rate from 7su/ha to wintering 11su/ha today. More evidence change was positive.

Grass grub is a big problem for all local farms and Travis' previous experience with fescue presented a logical solution. Instead of winter cropping, each year a selection of paddocks are direct-drilled straight into fescue after spraying glyphosate over existing pasture.

> Landcare Research estimates the benefit:cost ratio for agricultural weeds is 155:1, for every dollar spent, there's a benefit of $155.00. With ragwort, it's estimated at 860:1 – Vince Payne.

A couple of years ago, out of curiosity, Travis trialled two litres/ha of glyphosate as opposed to the recommended 4-5l/ha, resulting in a slightly slower die-off but the same outcome. This action fits with a desire at Kepler to reduce pesticides and other agrichemicals. This, in combination with longer grass, supports greater biodiversity which in turn helps keep pests in check naturally.

Now half of Kepler is in tall fescue that outperforms ryegrass on monthly growth rates whether wet or dry. Its dense roots reduce grass grub damage, unlike ryegrass, and buffers pasture production from wet or dry conditions and pugging. Due to fescue's upright growth habit, clover growth is supported rather than smothered. Today there is increasing uptake of fescue on Pāmu properties resulting from the experiences at Kepler.

Fescue ushered in changes in grazing management. Set stocking only occurs at lambing, and half the ewes come down to rotate on fescue flats once lambs are mobile. This removes pressure from ewes remaining in set-stocked country, allowing them to also thrive. Ewes and lambs on fescue flats pick the best pasture, leaving rougher feed for cattle to follow. During this time cattle maintain pasture quality for lactating ewes.

With no early lambing or hogget mating, all resources target mainline performance. They source replacement ewe lambs from other Pāmu properties, which enhances a full terminal programme. Rams are sourced from Kepler's own FocusPrime stud, which supplies Pāmu properties and beyond.

After weaning, focus shifts to cattle on 16-day recoveries to fast-track beasts for slaughter before winter. Pasture residuals are left high to ensure more growth and build insurance against dry autumns. Cattle suit this system well, being able to successfully harvest and convert coarser feed into weight gain.

> If methane links to kgDM eaten, then optimising gains in first year and having animals gone is a must – Adrian and Pauline Ball.

With changes in stock policy, beef production is now a key driver of Kepler's profitability. Animals with the shortest stay are most profitable, so they target weaning 250-300kg calves at 200 days to ensure finishing before their second winter (18 months) at 300kg cwt.

To highlight their success, a line of 150 homebred yearling steers processed in one day during May 2024 achieved 85% Eating Quality (EQ), demonstrating animals grew without stress or any growth checks, while scoring highly for desirable eating traits including tenderness, pH and marbling.

Kepler continues to evolve. As of winter 2024, it ran 3,300 commercial ewes with 600 purchased ewe lambs, a FocusPrime terminal sheep stud consisting of 1,300 fully recorded ewes, 500 ewe lamb replacements, and 600 ram hoggets wintered. Also carried are 600 in-calf cows which are currently part of an INZB progeny trial, and 1,400 head of trading cattle. For the past two years over 2,000 head of cattle have wintered on all-grass.

Travis Leslie's journey at Kepler Farm testifies how a positive management culture can transform a marginal block into a highly successful and resilient property. It demonstrates how incorporating resilience and flexibility creates better management outcomes than targeting just production alone.

Eco-profit Observations

- Reducing ewes by 500 maintained lambing percentage.
- Introducing tall fescue lifted grass production.
- Selling all lambs as either stores or primes at weaning switches all focus to cattle production.
- Reducing winter crop for all-grass wintering produced many benefits for all livestock classes.
- All-grass winter was better for ewes' teeth, required less labour, and resulted in more paddocks in spring, reduced costs by $100,000 and lifted overwintering carrying capacity.

- Reducing glyphosate application rates resulted in slower die-off but the same outcome.
- Taller grass supports biodiversity, which keeps pests in check naturally
- Tall fescue consistently outperforms perennial ryegrass, reduces grass grub impacts, promotes clover and reduces pugging.
- Set-stocking only occurs at lambing, otherwise all livestock classes are rotated.
- Source replacement lambs from other Pāmu properties allowing them to focus on main line performance.
- Cattle performance targets slaughter before second winter with high Eating Quality.
- Incorporating resilience and flexibility creates better outcomes than focusing on production alone.

17. Forbes and Angus Cameron - Measurement is Key

When it comes to mothering ability, "What you cull a cow for, give a ewe two ticks!" says Forbes Cameron. With 40 years of breeding sheep and cattle under his belt, he and son Angus spend a lot of time looking for outliers, including intelligent behaviour.

There's a catch in the breeding game; any outlier must accommodate several traits. Forbes recalls culling New Zealand Wool Board's Growbulk flock, a project to develop high bulk wool from Texel, Poll Dorset and Romney crosses. Only four farms across New Zealand had them.

Forbes found their culling impractical. For example, one ewe kept for wool bulk produced a single lamb in six seasons. Single trait selection creates undesirable outcomes, so they culled over 40% for lack of multiple benefits. Growbulk are highly prone to the Barbers Pole worm due to Texel genes.

On this basis, Forbes culled his highest marbled ram hogget because it had dags. Marbling premiums would never cover its health costs. What if premiums disappear or aren't paid?

On their 1,400ha hill country property, Ngaputahi Station, near Ashhurst, Manawatu, they've developed a low-input, highly fecund Romney flock. Due to the wide range of traits they select for, it's taken a while. On the way they won the Glammies twice while their stud beef won Steak of Origin.

They've collected their ram hoggets' faecal egg counts for 33 years. For the first 10-12 years, they struggled to link growth rates to faecal egg counts (FEC) but later started connecting growth rates and worm resistance with animals less prone to dags. In 1999 they crutched 70% of lambs, now in 2024, it's closer to 5%. Saliva testing identifies animals that secrete an antibody that stops worms attaching to intestine walls and which heals gut linings.

Many said this long-term measuring of multiple traits was too difficult. The Camerons did it old school by recording 4,500 individual lamb weights at birth every lambing for over 30 years. At the same time, they scored ewes on mothering and constitutional traits.

After weaning and mating, all commercial and recorded ewes are put in one mob, around 4,000, because this pressure creates culls. No dipping, no drenching. After weaning, ewes are grazed hard to get a worm challenge, usually in mobs of 1,500.

Forbes researched flystrike in the 1980s and now genetic records deliver valuable insights. A single set of triplets will get fly blown while hundreds of lambs in the same mob remain unaffected. A bad fly event in March 2006 resulted in more than 30 cases in one mob, 18 from one ram. It happens with lice and dermatitis. Remove those prone and keep those that don't react. Always treat those susceptible and sell them.

As ewe body condition scoring is no indicator for fly, Forbes suggests an enterprise for entrepreneurs; breeding sniffer dogs to detect fly-prone animals. Afterall, dogs sniff out explosives, drugs, human emotions and warn owners of imminent seizures. It's an example employing the 'dogginess' of dogs to create an unfair advantage. How many other ways can the intelligence of working dogs help farmers?

The Camerons don't cull on age or teeth so have eight- and nine-year-old ewes with no dags. Clients wanting flock longevity will ask for rams from ewes older than five years. The soil at Ngaputahi is sandy so if an animal can perform there, it should do well elsewhere.

Forbes and Angus don't feed sheep minerals because relying on supplements to get performance perpetuates a deficiency that increases costs. It's no different to drench capsules. That technology only breeds drench-resistant worms, and ends up costing farmers more.

Animals do not seek shade, a phenomenon they believe is due to not using nitrogen fertiliser – Hamish Browne and Robyn Budd.

They're not interested in big ewes so produced a smaller frame. Ewes at 53kg liveweight will scan 200%, and at 60kg scan 236%. Highly fecund means even in a bad year ewes have a high chance of getting pregnant.

Hence, clients will buy hoggets because they scan 150%. Hogget ability to get in lamb allows the Camerons to command a 20% premium. As their stud size increases, they finish fewer commercial lambs so premiums make up for that.

Back in 1983 they started measuring an easy trait to improve sheep quality, the eye muscle on the twelfth rib, and then in the late 2000s moved to marbling in cattle and later to intramuscular fat in sheep. Now farmers are getting paid

> Today Atkins pays farmers to do their own drafting and for keeping to a production schedule. This is an important industry lesson, rewarding farmers for their preparation while the whole supply chain benefits – Mark Warren.

high premiums because abattoirs measure these traits. Marbling in cattle lifts returns by $12,000 per truck and trailer unit. They once bought a facial eczema-resistant ram and found it marbled highly. What value did the breeder miss because it was never measured?

A breeding unit is about numbers. As the saying goes, two stores always beat a prime for income, particularly if it hasn't cost anything up to weaning. If a 60kg ewe weans double her liveweight with quads, that's efficiency because the ewe is the mother, not the farmer. While not yet consistent, it's been done at Ngaputahi.

How could this be possible? An interesting observation reveals sheep intelligence. At least one quad-bearing ewe planted lamb pairs either side of a paddock. This means she can fully feed each pair while grazing on journeys between them.

A problematic issue with modern sires is longevity. Bulls and terminal rams will not last if overfed. One cause is agents preferring large, fat animals because commissions are weight-based. Forbes recalls Australians found 20-month bulls' ideal liveweight to be 550kg, otherwise they become too heavy for their age and damage their feet and pasterns.

Over 40 years, the Camerons have created many genetic insights to reduce costs and risks and to lift premiums. If farmers want an unfair advantage, the Camerons prove observation and measurement is essential for results.

Eco-profit Observations

- Over 40 years they've developed a low-input, highly fecund Romney flock.
- Single-trait selection creates undesirable outcomes. Forbes culled his highest-marbled ram hogget because it had dags.
- In 1999 they crutched 70% of lambs. Now, in 2024, it's closer to 5%.
- Saliva-testing identifies animals secreting an antibody that stops worms attaching to intestine walls and which heals gut linings.
- Some animal health issues are genetic. A single set of triplets will get fly-blown while hundreds of lambs in the same mob remain unaffected. It happens with lice and dermatitis too.

- Forbes suggests an enterprise for entrepreneurs; breeding sniffer dogs to detect fly-susceptible animals.
- Clients wanting flock longevity will ask for rams from ewes older than five years.
- Don't feed sheep minerals because relying on supplements to get performance perpetuates a deficiency that lifts costs. It's no different to drench capsules.
- Ewes at 53kg liveweight will scan 200%, and at 60kg scan 236%.
- Hoggets' ability to get in lamb allows the Camerons to command a 20% premium.
- If a 60kg ewe weans double her liveweight with quads, that's efficiency because the ewe is mother, not the farmer.
- Ewe intelligence: Planting a pair of lambs either side of a paddock allows ewe with quads to fully feed each pair while grazing on journeys between them.
- Bulls and terminal rams will not last if overfed. They become too heavy for their age and damage their feet and pasterns.

18. Tom Hargreaves - Breeding Priorities Key

Essentially, beef cow economics is based on the cows eating feed of little value to other animals. The key is matching genetics with environment to optimise calf growth and production. This backgrounds Kakahu Stud's success.

The Hargreaves, farming near Geraldine, South Canterbury, are into their fourth generation. Twenty years ago, Tom left his Melbourne architecture business to come home. His father Gerald told him not to come home before 30, so after a year at Lincoln he saddled up for the European polo circuit before settling on a degree in architecture.

Kakahu is 1,600ha (along with a neighbouring 200ha dairy farm) running dairy, beef and sheep, plus Angus, Charolais and Changus studs. There are 450 Angus cows, more than 200 sale bulls and 250 heifers, plus 90 Charolais cows and their progeny as well as 60 Changus heifers and 2,300 crossbred ewes. Next door there's 450 dairy cows.

The beef studs centre the operation. Tom's grandfather Kenneth started the Angus stud in 1954 and Gerald took over in the late 70s. At that time a trip to North America inspired the use of artificial insemination instead of natural mating and they've steadily imported large quantities of Angus semen since. Sourcing bloodlines from colder, drier climates creates a resilience and performance Kakahu is known for.

The Charolais stud started in 2006 to give clients more cross-breeding options and now the main benefit is calving ease for dairy cows. They almost cashed it up but then one bull sold to Xcell Breeding Services, Kakahu Gerry, fathered 380,000 calves. They no longer sell bulls for semen, instead they are paid a semen royalty.

The Changus stud emerged in 2017 with Tom's cousin Hamish Orbell at Clayton Station. They aim to breed first-cross (Angus-Charolais) bulls for dairy farmers, specifically targeting calving ease with high growth. Bulls are sold as yearlings. The intention is breeding a white Angus that has all the benefits of Angus and Charolais combined for semen collection, similar to Kakahu's Charolais stud.

Strategically, Tom's focus is creating a better income balance over multiple income streams. Historically, Kakahu generated two cheques: bulls 80% and lambs 20%. Last year's bull sale cheque was 52% of total income so he wants that down to 30% to reduce risk. In this, he's mindful of succession.

It takes dedication to run a stud. While the family will never sell the farm, the next generation may sell the stud. Currently, such a shock would cripple the whole business so he's creating more flexibility as he prepares for that possibility. The dairy farm is part of the succession plan for son and daughter.

Tactically, Tom employs a number of savvy business insights. For example, how do you pump out more progeny without growing the herd. Surrogacy. In the cattle stud game it's buying a dairy farm. Dairy cows carry stud bull embryos to term while squeezing out milk during gestation. It's the best of both worlds, improving herd genetics without buying more beef cows and beef country while producing continuous cashflow from milk.

Furthermore, by flushing their best beef cows, they're not limiting each one to raise a single calf each year. That lifts herd genetic gain and production statistics faster. However, only 45% of embryos stick, so efficiency has a long way to go. Beef-cross heifers prove best at taking embryos to full-term, not dairy cows. That said, 100 top cows carry embryos that are 100% Angus or Charolais calves.

The potential of embryos puts pressure on other livestock classes. For example, because it essentially costs the same growing out or buying in replacements, Tom is considering dropping hoggets and buying two-tooths to feed more cows carrying embryos, despite crossbred hoggets achieving 109% lambing.

There's a paradox stud breeders juggle: Every commercial operator wants cows that lay down fats but processors don't want back fat on carcasses as cutting it off adds costs and time. For buyers it means focusing on two important traits: feed efficiency and cow liveweight. When buying a bull look at the cow.

> The move to more meat production has lifted wool micron which has been as low as 13.8 but is now around 16.5 – Paul and Prue Ensor.

A smaller animal with higher weight gain is always more efficient than heavy cattle, especially on hill country. Mature cow weight (MCT) and 400-day (400-D) liveweight should be similar so progeny must perform. Adding easy calving to the mix reduces client stress further. Ideally, Kakahu clients are getting more gain from every mouthful of feed.

An important breeding realisation is that the best-looking animals may not generate the most revenue, hence measurement (EBVs) comes before form and efficiency is forefront when breeding valued sires.

Tom believes carcass weight and meat quality are the priority. Rib is key, rump has less relevance. The hardest part is carcass quality so eye muscle scans focus on that. DNA testing carcasses finds which sire lines are yielding. Carcass weight matters most as farmers aren't paid well enough for marbling. Kakahu aims for carcass yields of 58% compared to the national average of 54%. That extra adds to bottom lines over time without doing too much extra.

Jersey cows are known for their butterfat and the Heald's' herd delivers 7%, compared to Friesians at 3.5%. This is significant – Russell and Charlotte Heald.

In a world where artificial meats are tipped to be the next big thing, Tom notes customers want flavour and tenderness, so delivering that experience is crucial. It requires low pH, good meat and fat colour, and above-average IMF. Some Kakahu clients have more than 80% of their cattle achieving Silver Fern Farms' Eating Quality (EQ) grade compared to the national average of 30%.

Tom runs a sharp pencil over operations at Kakahu. While he's targeted specific metrics and ways to expedite genetic progress across all three studs, his overall aim is strengthening family resilience to business shocks. After all, as he states, "Time takes care of debt."

Eco-profit Observations

- Match genetics with environment to optimise calf growth and production.
- Sourcing bloodlines from colder, drier climates creates resilience and performance.
- Started a Charolais stud for more client cross-breeding options with the main benefit being calving ease for dairy cows.
- Looking to future shock-proof the family farm by diversifying into businesses other than studs.

- Increase stud progeny with embryo transfers into the dairy herd. Means they can lift genetic merit faster as one cow can produce many offspring in a season while producing cashflow from milk production.
- The income-generating ability of surrogacy puts pressure on other classes of livestock to perform.
- When buying a bull look at the cow.
- A smaller animal with higher weight gain is always more efficient.
- Rib is key, rump has less relevance. The hardest part is carcass quality, so eye muscle scans focus on that.
- Customers want flavour and tenderness which requires low pH, good meat and fat colour, and above-average IMF.

19. Richard Thorp - Set-Stocking Success

Does set-stocking pay? It can, according to ex-rural banker Richard Thorp. With a five-year average EBIT of \$53/su and farm working expenses of half that, it's allowing him to continue doing what makes most young farmers cringe - the same management as his father. Now in the twilight of his career, new challenges have ignited his passion for change.

His grandfather created the property overlooking Kopuatai Wetland in 1948 by buying several small dairy farms along the old Paeroa to Te Aroha road before Waikato's swamps were drained. The 520ha-effective hill property runs close to 6,200su comprising 200 mixed-age Angus cows plus in-calf heifers, another 330 head of non-reproductive cattle classes, 2,100 mixed-age Romdale ewes, 500 ewe hogget replacements, 50 Wiltshire rams and another 550 head of trading stock of various classes.

All paddocks were set up in the 1950s with an all-natural water supply. He has no stock water system to maintain. When dams and small creeks dry up, he keeps gates open for stock access to water. There are five sets of satellite yards for docking and other jobs with a laneway back to the main sheds and yards. The highest point is a trig station at 300 metres.

The property attracted publicity in 1972 when his father, Basil, was filmed for the TV show Country Calendar. Basil was a time and motion man inspired by Henry Ford. His focus was finding time efficiencies to juggle stock work with development fencing. He gravelled all laneways and built cattle stops instead of swinging gates. Satellite yards are situated atop rises as livestock prefer to run up hills.

Fifty years later when Country Calendar rang wanting to show the management changes since he took over from his father, Richard said there were none. The filmmakers simply spliced black and white footage of his father with Richard's modern-day activities.

Richard's no hayseed, though. Arriving home in 1997, he worked for his father for several years. He'd seen many production systems while a career banker and thought he could crank up intensity. However, once he began growing crops and rotational grazing, he found he couldn't do everything himself. He came to realise Basil's way was how one man could work efficiently and cover all the ground.

> However, he acknowledges fodder crops cause many more animal health issues than all-grass wintering and luckily those costs fall on dairy farm gross margins – Will Perriam-Lampp.

He says his 10 years in banking revealed it to be a culture of competition - celebrating who was going bigger and faster. Yet he found some of the best farmers didn't need financial programmes telling them what to do. They were unassuming, humble and with impressive farm surpluses. That's all that mattered. Upon leaving the bank, the first thing he got rid of was his watch.

After mixing with indebted clients with big overdrafts, he found himself in that situation taking over the farm. When farming to the wire, there's less room to move. Over the last 10 years he has made big gains from profitable sheep and beef breeding, not trading, and has paid off debt principal. When prices are up life is easier.

Now he's not so focused on measuring production and doesn't worry about how much fertiliser he's putting on, usually applying 100 tonne of super 10 or less. Taxes are his biggest bills, meaning there's no incentive to produce.

Keeping things simple is about sanity. Ewes and cattle swap paddocks every six months, triggered by weaning lambs late November/December and calves late May/June. Weaning lambs coincides with destocking into summer with 800 stores and 200 primes leaving the property. Anything left is kept and sold as winter-finished. Weaned calves are fattened and sold into the local market, heifers at two years, steers at three. This simple grazing routine allows him to keep on top of maintenance and weeds.

Richard doesn't like destocking or altering set patterns. Last autumn (2023), it didn't rain from January to June. That five-month dry hurt. The little rain in April caused facial eczema and sick lambs and ewes. When combined with a poor tupping, lambing tumbled to 85%. However, there's enough fat in the system to push through ups and downs.

With set-stocking, fertility shifts from hillsides to hill tops and valley floors. Richard notes that while his rolling hill pastures are very diverse and have good clover content, steeper hillsides have good populations of ratstail (*Sporobolus africanus*) and other low-fertility species. Other issues include thistles after drought or dry spells when pastures are eaten very low, though livestock will also eat thistles if wilted. Being non-brittle, bare soil from overgrazing covers over quickly.

> They're considering expanding to 40-50ha but unsure if they want the hassles of going bigger, such as employing staff – Ross and Karen Potter.

Labour is a huge cost and like his father, Richard admits not being great with staff. It's hard to find good casual staff but Cam Clayton, who recently returned to Paeroa, is very capable during busy periods. Richard's recently leased 150 acres to him to reduce workload while mindful of how hard it is getting started. If the leasee has little equity in livestock, it's even harder.

That said, he's finding it hard to let go because of his family history of being good curators with a tidy farm and a reputation for good stock. While the shift to Wiltshires has reduced shearing costs and labour, Cam challenged Richard about meat production and lambing percentage which got him looking at the figures again.

While Basil was all about efficiency, Richard's corporate career opened his mind to other paths. He spends time hunting, climbing and mountain biking and enjoys flexibility to help out his children. He's moved beyond the pressure to prove anything and that's all that matters.

Eco-profit Observations

- A five-year average EBIT of $53/su and farm working expenses of half that, allows Richard to continue his father's practices.
- No stock water system to maintain, all-natural water. During dry times, opening gates provides livestock access to water.
- Five sets of satellite yards for docking and other jobs with a laneway back to the main sheds and yards.
- Father, Basil, inspired by Henry Ford, installed cattle stops along main laneway to hasten travel times.
- Satellite yards are situated atop rises as livestock prefer to run up hills.
- Once growing crops and rotational grazing, Richard found he couldn't do everything himself.
- As a rural banker, he found some of the best farmers didn't need financial programmes telling them what to do.
- Taxes are his biggest bills, therefore no incentive to produce.

- Simple grazing routine allows him to keep on top of maintenance and weeds.
- Rolling hill pastures are very diverse and have good clover content, steeper hillsides have good populations of ratstail and other low fertility species.
- Richard finds it hard to let go because of his family history of being good curators with a tidy farm and a reputation for good stock.
- A shift to Wiltshires has reduced shearing costs and labour, but being challenged about meat production and lambing percentage has got him looking at figures again.

20. CONSULTANTS BREED COSTS

These analyses clearly show that farm viability depends more on minimising losses than maximising production, and it is these accumulated losses which threaten farm business survival and growth.
Tim Hutchings, Australian agricultural economist (2013). Financial Risk on Dryland Farms in South-Eastern Australia. PhD thesis.

Allan Nation's expertise involved looking at principles of big business success and then finding farmers and ranchers applying them inside the farm gate.

For example, in Creating a Family Business he explained how business traders stripped equity to finance other deals. Wall Street managers spent their time looking past accounting fiction, like depreciation, trying to understand a business' true cash flow position, because upon buying the company they would load it up with all the debt its cash flow could sustain, transfer the loan elsewhere and then quickly sell off the company with its debt or even bankrupt it. This created opportunities to secure funds (equity) without the strings of banks or investors. Nation said this was why small business operators should never take advice from Wall Street managers because they removed themselves from the responsibilities of debt.

Do farmers do the same thing? Absolutely. If a low-input property comes onto the market it'll be quickly snapped up, even at a premium, because buyers believe its performance will increase when "farmed properly". Production is cranked up to drain its natural capital (equity). Once gone, its ability to buffer downturns evaporates along with ROI, so it's put up for sale and promoted with inflated production based on inputs. This behaviour showcases how farmers and their advisers, particularly banks, strip natural capital for short-term financial gain thereby ignoring the resilience values ecosystem services provide.

But surprisingly, many investment industry giants would argue this asset-stripping is not best-practice. Investors like John Bogle and Warren Buffett write a different view to the cut and thrust of Wall Street trading. Anybody reading Buffett knows that when buying a company his focus is cash flow and relationships with customers. Customers are the building blocks of its financial ecosystem; without them everything collapses quickly. He doesn't drain staff or books for a quick gain. He knows bottom-lines aren't just on paper.

Price is what you pay, value is what you get and Buffett observes investing behaviour hasn't changed much over the time he's been doing it. The dumbest reason to buy something is a rising price. Emotion rather than logic drives trades. Bogle is also critical by stating nobody is interested in looking under the hood of their purchases. Ever notice university graduates buying sires solely based on breeding index values? They're unable to assess a standing animal.

> No other global celebrity championing the environment comes close to carbon positive because they cannot relate these principles to livestock farming. And few, if any, would copy his diet – Why Bear Grylls should Spearhead Red Meat.

Yet how many farmers and their professional advisers understand the relationships between the four ecosystem processes (water and mineral cycles, solar energy flow, and behaviour of biodiversity), that are the basis of farm production? Strip one away and the whole collapses. Farmers (investors) get in trouble when buying low-input properties by thinking they can perpetuate elevated production (through purchased inputs like fertiliser) only to find a lack of natural capital fails to buffer rises in debt-servicing, lower commodity prices or higher input costs.

Impatience blinds investors to the principle of compounding. In The Little Book of Common Sense Investing, John Bogle demonstrates compounding interest based on his firm's financial products. A single dollar at 7% return reaches $2 in a decade, doubles to $4 in two decades, $7.50 in three decades, $15 in four decades and $30 in five decades. Nature operates the same way, but unfortunately banks are every bit as greedy as Wall Street traders. The industry is solely focused on short-term gain, and not the farmer.

> Why reduce debt? Pita observes there's not much else that delivers an 8% return with cash surpluses – Debt Creates Futures.

Note agricultural schools are no different to business schools; they reward complicated behaviour more than simple behaviour, even though simple is more effective. Students are no longer taught Occam's Razor - when multiple solutions to a problem exist, choose the simplest. It's usually grazing management, what could be termed Occam's Grazor. Changing mob numbers, size and density, as well as grazing and recovery periods using an overhead, the shepherd's wage, often has a greater ROI than commercial interventions. This is not the focus of industrial professionals because using natural processes like grazing cuts them out of the picture.

The problem is, farmers confuse simple with easy. It's easy to get professionals to sort problems. Just as Buffett and Bogle observe, professionals offering complicated solutions crank up their fees to profit from investor ignorance, impatience and greed. Farming is no different. Complicated operations usually have an army of advisers. So, any solutions?

To remove emotion from trading, Bogle developed Vanguard's indexed funds, allowing investors a share of the whole market. Why hunt for a needle when purchasing the whole haystack is more effective, especially if holding over the long run. This step also eliminates the siphoning off of fees by professionals and their illusionary short-term profits. Similarly, shareholders of Berkshire Hathaway have spent 60 years continuously reinvesting instead of drawing dividends.

Bogle states financial alchemists can't turn lead, copper and iron into gold, they can't outperform the market over long periods. Every agricultural scientist knows season trumps treatment. In farming, no agrichemical or fertility product sold outperforms itself in a poor season compared to a good one. When farmers realise this, they're in a much better position to focus on where investment needs to be made. That investment relies on improving resilience.

The closest we can get to this in farming is focusing on soil carbon to store moisture and fertility for tough times. Farm profits are largely related to soil organic matter. The first 30 years of cereal cropping on the Canterbury plains achieved record yields without manure or fertiliser applications.

Just like sharemarkets, the secret to farming isn't maximising short-term returns but minimising accumulated losses by keeping costs down. Unfortunately, that requires time and conflicts with impatient bank managers and shareholders.

Eco-profit Observations

- Small business operators should never take advice from Wall Street managers because they removed themselves from the responsibilities of debt.

- Farmers and their advisers, particularly banks, strip natural capital for short-term financial gain thereby ignoring the resilience values ecosystem services provide.
- Investors like John Bogle and Warren Buffett write a different view to the cut and thrust of Wall Street trading.
- The dumbest reason to buy something is a rising price. Who looks under the hood of their purchases?
- A lack of natural capital fails to buffer rises in debt-servicing, lower commodity prices or higher input costs.
- Impatience blinds investors to the principle of compounding.
- Students are no longer taught Occam's Razor - when multiple solutions to a problem exist, choose the simplest.
- Just as Buffett and Bogle observe, professionals offering complicated solutions crank up their fees to profit from investor ignorance, impatience and greed.
- Why hunt for a needle when purchasing the whole haystack is more effective, especially if holding over the long run.
- In farming, no agrichemical or fertility product sold outperforms itself in a poor season compared to a good one.
- Farm profits are largely related to soil organic matter.

21. Michael and Catherine Reilly - Mixing with Like-minded

Michael Reilly's most vivid childhood memories are of the freedom of open spaces and how soothing it felt to squeeze milk from his father's house cow.

Two decades later, living near Sydney's northern beaches, the confines of urban living left him city-sick. He kept chickens, tended a vegetable garden and went on farm-stay holidays, but it wasn't enough. It was time to move.

Michael's wife, Catherine, had family in Hawke's Bay. They cut ties with their Australian home and moved to New Zealand to be closer to her family. They found an ideal 11ha lifestyle block in hills south of Waipukurau; no snakes or spiders. The next few years Michael floated between his family's Sydney construction business and various local rural Hawke's Bay jobs. The neighbouring farm had been on and off the market for several years, so they made an offer.

They bought the 170ha property in 2016 and soon the stars in their eyes were eclipsed by their bank's business coach. They did everything by the book. Their first adviser set up a great plan with cattle to get the loan only for the next one to throw it out and draw up another for trading lambs. It looked much better financially and with all the assurance and support of professionals, how could they fail?

> They chose to eliminate fertiliser. Each year they waited to see if industry advice was correct and pasture production would collapse. Some 35 years later they're still waiting – Geoff and Barbara Croker.

Michael realises now they were thinking the wrong way. Their new plan involved spraying crops, reseeding pastures and heavy fertilising. Usually, these ideas are endorsed by well-established farmers looking to transplant their success elsewhere long after their financial pressures have eased. After his experience in his family's business, he wondered whether the integrity of farm consultants and their products could be compared with that demanded in construction.

The first 12 months were characterised by good trading advice for buying and selling, allowing Michael to build up his knowledge of how livestock markets worked. However, while money was good, start-up costs pushed through to year two when it became clear lamb trading profits weren't covering pastoral development costs.

Upon being told by a consultant to spray his crops with insecticide, Michael gave contractors a list of paddocks but not the whole crop. He was curious if spraying made a difference. It did. It clearly made plants weak and spindly; their leaves became brittle. Even though bugs were still present in unsprayed areas, their numbers declined and plants were stronger and responded better to weather. From that moment Michael started questioning industry advice.

Furthermore, it dawned on Michael and Catherine that while they happily started off with conventional farming, they didn't want to stay there. They didn't want agrichemicals. They were eating organic food but not their own produce. What example were they setting for their children?

With meat prices forecast to fall, they saw issues continuing with crop and pasture renewal. Crop ground had dry patches that never grew back and eroding soil. Michael was working 35-hour weeks for neighbours and fencing contractors plus his own farm. He crunched a budget showing a $20,000 deficit.

His adviser's solution was to simply work more hours off-farm. That attitude hit hard because there was no curiosity to look for solutions. Till then their adviser had been very helpful but knowledge beyond his own money-making techniques was limited. Breaking this relationship was essential to move forward.

Lamb trading proved too expensive so they began connecting with other farmers. Michael started mixing with like-minded locals, Phyllis Tichinin, Murray Cammock and Will Foley, and realised these people weren't country bumpkins. They didn't see brick walls like his neighbours and introduced him to many farmers doing things differently in the community. He began building confidence by adapting their ideas.

He and Catherine moved to safer profits with less costs, still trading lambs but the bank prevented them starting a beef breeding unit as it wanted consistent cashflow. When Covid hit they sold ewes below cost and trucked out dairy heifers, but had no grass to fatten anything so lost six months income. They rotated everything else in one mob which stimulated pastures to surge four weeks earlier than their neighbours.

Michael could see their mortgage being paid in five years but then the worst drought in 80 years hit. One day, pulling lambs from a muddy dam, Catherine questioned whether this was what they wanted to do. Michael began wondering if he was addicted to the farm. He hadn't asked his family how they thought things were going. If it didn't fit with them, then did it work for him? With growing teenagers, did they need to be closer to town?

Spring proved a cracker, pushing grass along just with grazing management. With grass growth tracking well after the drought, they did more fencing and gates and put water troughs out back. Within two weeks of installing the last trough, the property was on the market. People couldn't believe what pasture they saw and were amazed what was growing despite no fertiliser in two years. It was all down to grazing management skill.

> But in a location prohibiting synthetic fertilisers and without new pastures, how did they produce award-winning meat? – Colin and Keryn Brown

Statistics for business failures by year five are high, but many like the Reillys don't collapse financially. It's simply the stress to align everything for success that takes a toll. The experience didn't put them off farming and they have clear priorities to start again.

Doing a Holistic Management Course clarified their family's lifestyle and business journey and the corrective actions needed to maintain it. Now, they would gain organic certification and secure market outlets earlier, as well as source better genetics to suit their property and management. Most importantly, they would build a community to surround themselves with like-minded people for support.

Eco-profit Observations

- Banks focus on cash flow at the expense of profit.
- Began questioning integrity of farm professionals.
- Found out early revenue wasn't covering enterprise start-up costs.
- Observations from unsprayed crops lead to questioning advice.
- Farm practice was conflicting with family values.
- Professionals had no curiosity to look beyond their own knowledge.
- Find like-minded people to confidently try ideas with safer profits.

- Is family enjoyment of their farming experience a sign to continue?
- Used grazing management to stimulate pasture growth from drought without fertiliser to lift property value.
- Experience has them secure for starting their next go at farming.

22. Shane Armstrong - Stepping Off the Treadmill

Opotiki dairy farmer Shane Armstrong increased fertiliser inputs and supplements, all driven by LIC genetics, only to crash. A management change produced four environment awards in the 2023 regional competition, almost half the prizes. But no mention was made that he'd ditched the sponsor's fertiliser products and company shares several years earlier.

What increases pressures on young farmers are promises that never materialise. Despite flying on fertiliser at high expense, grass only grew where cows urinated, especially on hills. Fertiliser reps repeatedly advised increasing potassium but no change eventuated other than higher costs. But it wasn't just fertiliser costs that were climbing.

Brett has observed livestock behaviour changes before production improves. Firstly, anxiety reduces and animals become more relaxed – Brett Petersen.

Downer cows dropped like flies, with retained placentas, sore feet, stone bruises and Johnes disease. It's no wonder farmers are plagued with mental health problems. The standard "more-on" approach to problems, accelerated others. This treadmill was driving Shane crazy.

Then in a lifestyle magazine he found Greg Hart from Mangarara Station and his list of books to read, which introduced him to Joel Salatin. Reading Joel's books, he discovered another way of profiting from farming without intensive production.

He changed consultants to provide him with soil observation skills and the courage to wean himself off industry products. A Holistic Management course strengthened business skills by linking profit to nature through new insights. For example, bulls now graze in odd corners for finishing, turning these problem areas profitable.

The journey to low-input systems isn't easy because everything's geared the other way. Production dropped from 1,100kgMS/ha to 610, along the way briefly holding at 800 which is his target. Moving to once-a-day-milking (OAD) usually drops production by 10-15%. While cows keep better body condition, first-calvers struggle to milk well and this hampers production, culling depth and herd progress.

However, there's no more calving troubles, feet issues and Johnes. Breeding is much easier as there's no need to check cows for infections, no cidrs, and mating takes seven weeks instead of 12 so there's no extra feeding required. He has 7-8% empties with MUNs at 10-20. Artificial insemination lasts three weeks with the bulls out for another 25 days.

But mastitis is a major concern. Shane suspects years of using high breeding worth LIC bulls produces udders barely lasting four years. They sag, lose shape and become unmilkable. A shift from Kiwi-cross to Jerseys is improving udder structure and shape, and this is further strengthened by picking dams with longevity for breeding his own bulls. Introducing the hardier Viking Red helps but they start drying off after Christmas on OAD.

Despite lower production and higher farm working expenses (FWE) at $5.20/kgMS, his profit is still above the district average because FWE have climbed faster for conventional farms. He enjoys drawing an organic premium, has great staff retention, healthier cows and time to chase his passion of endurance running which takes him all over New Zealand. All this from 280 cows on a 150ha platform.

> Unlike many of his mates, he went small.□ Now he still enjoys milking cows, with no staff and no burnout – Graham Smith.□

Peer pressure from award-winning farmers and local regional council staff got him into the local environment awards. That experience left him thinking such events are largely to satisfy public perception rather than help farmers. The initial judges were critical, stating that without a soil test for several years, he was mining his soil. In frustration Shane threw his spade into the ground and dug a hole. He'd split a chicory root and surrounding it was soil carbon centimetres thick against the yellow clay. "This is what I'm doing," he said and out came the judges' phones as if they'd never seen soil carbon before.

The final feedback from judges only focused on lifting monitoring efforts for future compliance, nothing practical such as specific trees to improve landscape function, how grazing management improves soil health, nor anything linking soil health to animal health and profit. Frustratingly, other farmers are celebrated as sustainable despite their maize crops flooding waterways with silt. Dropping maize is what Shane credits for improving soil health because it reduced so many compaction and runoff issues.

Even today, when groups led by industry professionals visit his property, none have any soil knowledge. They look at numbers on paper instead of the soil itself. Their idea of water quality is constantly tied to planting trees and to riparian fencing instead of grazing practice and soil health.

For Shane, also gone is ripping up and sowing down pasture. He observes deferring grazing and fallowing pasture are producing swards suited to his management and environment. Longer pasture has almost eliminated green bristle grass weed. Ryegrass also declines, but it continues to push through paspalum setting up for winter. Pastures are still pumping out 15,000kgDM/ha under organics.

However, grazing management is a skill proving tricky to master. Reaching the optimal three leaves per tiller is challenging but it's the cheapest way to ride the wave of growth into summer and lengthen the round. By Christmas, the aim is to reach 40 days recovery. Then at the beginning of February, as summer grass kicks in, he'll cull empties and use supplements to push recovery out to 50 days.

It's a tricky balance going into winter as often there's a lull in growth over late March into April as cooler season plants gear up and warmer season plants wind down. With Opotiki's hot and humid summers, he's noticed planting more trees is lengthening cooler season pasture production.

At least the next generation is already convinced. Young son Cooper is well skilled on grazing management. When walking into paddocks he yells, "Grass not ready yet Dad. Too many square tips!"

Eco-profit Observations

- Grass only grew where cows urinated but reps repeatedly advised increasing potassium resulting in no change other than lifting costs.
- The standard "more-on" approach to problems accelerated others, mostly animal health problems.
- Changed consultants to provide him with soil observation skills and courage to wean off industry products.
- Moving to once-a-day milking, cows keep better body condition but first-calvers struggle to milk well which hampers production, culling depth and herd progress, yet no more calving troubles, feet issues and Johnes.
- Shift from Kiwi-cross to Jerseys is improving udder structure and shape.

- Hardier Viking Reds help with udder structure but they start drying off after Christmas on once-a-day.
- Despite lower production and higher farm working expenses, profit is still above the district average.
- Environmental award judges accused him of mining his soil until he dug a hole in the ground to demonstrate soil development.
- Final feedback from judges only focused on lifting monitoring efforts for future compliance, nothing practical such as linking soil health to animal health and profit.
- Industry professionals' idea of water quality is constantly tied to planting trees and riparian fencing instead of grazing practice and soil health.
- No more pasture development, instead deferring grazing and fallowing pasture are producing swards suited to his management and environment.
- Grazing management is a skill proving tricky to master as reaching optimal three leaves per tiller is the cheapest way to ride the wave of growth into summer.
- Planting more trees is lengthening cooler season pasture production due to shade.

23. Willie White - Organic Minimises Losses

What if the entire New Zealand dairy industry went organic? Ex-banker Willie White reckons that if this happened, conventional farms would make more than $1,000 a hectare extra.

He calculates this using his 2019 production figures and adds a 10% rise in costs. This shows, he says, it's not the payout that drives the difference between organic and conventional dairy farming.

This is why organic dairy farmers are chill every year, he says. While non-organic farmers laugh at them, he finds it amusing they're always the ones juggling more issues.

Willie worked at the Bank of New Zealand for 10 years. It taught him there were many ways of doing things. His two best clients were a dairy System 1 (no external supplements) and dairy System 5 (50% external supplements), demonstrating production systems didn't determine profitability. However, it did help him develop an aversion to high-input systems as when they went wrong, it was bad. Low-input guys were always doing better.

Now farming himself, near Waipukurau, Hawkes Bay, he runs a System 1 and milks a 1,000 cows on two platforms totalling 465ha. Both herds milk once-a-day. This changes production curves by flattening spring peaks with cows topping 1.7-1.8kgMS/day. Once-a-day milking sees production curves decline from February and pick up again in March where twice-a-day declines from December. Cows squeeze out around 350kgMS/head.

Being an ex-banker, ROI was the primer for looking after land. Therefore, he was happy to explore any system which compelled a closer look. While his family has been on the property since 1880 and converted to dairy in 1995, Willie grew up on the cropping farm next door so didn't have baggage coming from a dairy background.

Still, it took three to four years before he had the courage to switch. His conviction firmed when he dropped nitrogen and noticed it was "a bit of a scam" when farmers didn't always maximise a return on it. The traditional dairy industry system felt underwhelming to him, simply identifying what production a farmer wanted and then working backwards. He felt it made every year look good but never prepared farmers for the unexpected.

His tipping point came during John Meek's field day at Norsewood in 2016. That's when things fell into place. As he says casually, "I found a system I didn't know I was looking for." Overnight he made the decision to go organic and gained confidence from talking to a number of key people like Patoka organic dairy farmer John Kamp.

Now four years organic certified, farming is easier. Financial rewards during conversion are tough but farmers still get 45c/kgMS above conventional. Over the last five years Fonterra's organic milk price has been 30% above conventional. Now debt doesn't affect his sleep anymore. While his income fell 25% in the 2023/24 season, farm working expenses fell 50%.

Noticed pasture was always better with more clover growth in corners or areas where urea missed – Michael Mageira and Smita Apte.

One reason for fewer costs is the practice of deferred grazing instead of cropping. Replacing an input with a natural process like deferred grazing buffers the business against financial knocks. Benefits of lower input systems became clear when inflation rose after the covid pandemic.

He's tried cover crops but he now simply undersows bare pasture or uses hay to bring in seed to cover patchy paddocks. Willie used to buy 300 tonnes of annual fertiliser, now nothing. That creates pressure for fertiliser companies.

His current focus is minerals, particularly calcium, magnesium and sea minerals, as well as Ocean Base fish. What's been noticeable is available phosphate rising from 22ppm to 29ppm over four years, a rise of 30% despite not applying any.

Willie challenges farmers looking at regen to convert to organic instead. His experience is that stress drops in the first year of transition as finances sort themselves out and this eases further in the second year as everything slots into place.

With no low-hanging fruit in agricultural commodities, branding is one way for industry to get more value – Greg Crombie.

While organic is restrictive, having to work around issues simply means being organised and communicating better. Regen guys want a foot in each camp, blind to the contradiction that agrichemicals push ecosystems out of regen status, he says. They don't help farmers solve their problems.

One tool Willie tried was Halter electronic collars. Along with shed scratchies, Halter made it easier to identify cows in heat, especially by picking up cycling cows earlier. This meant they were serviced on that day instead of 24 hours later when visible symptoms were evident. It reduced wasted semen and inseminations.

With Halter, collaring takes a bit of time. Heifers are collared in their second winter by swapping collars with cull cows. This gets them trained well before calving. The system needs 14 days to get a baseline for each cow's data.

Initially he thought collars would make the farm a better place to work. Instead, staff spent too much time on their phones planning and lost touch with herd health and wellbeing. Collars resulted in even grazing residuals, but staff eventually lost the motivation to observe paddocks. Despite excellent service from Halter, collars made staff lazy resulting in poorer grazing control going into summer.

Willie remains critical of the traditional dairy industry. Farmers are guilty of doing the same thing over and over, few see the light, he says. Farm consultants only offer highly complex systems because any risk is with farmers. Farmers winning environmental awards are usually from high production systems, he says. Bankers are always production focused and few, if any, have any understanding of how ecology can minimise losses with livestock.

Fear of losing production is largely behind industry attitudes towards organics. As Willie White demonstrates, he's thriving despite going organic.

Eco-profit Observations

- Organic premiums are not responsible for organic dairy farming profits.
- Lower input systems have more flexibility when times are tough.
- OAD milking flattens spring production curves but declines late summer.
- Nitrogen fertiliser never guarantees maximum return on investment.
- Traditional dairy management of calculating desired production and working backwards never prepares farmers for the unexpected.
- Income fell 25% in the 2023/24 season, farm working expenses fell 50%.

- Practices deferred grazing instead of cropping which reduced inflation exposure during 2022/23 season.
- Phosphate rose 7ppm over four years despite cancelling 300t of fertiliser.
- Regen farmers blind to agrichemicals pushing ecosystem out of balance.
- Has found electric collars made staff lazy.
- Environment award winners are usually high production systems, seldom organic because of industry bias.

24. Kenneth and Rachel Short - Petrolhead Heaven

Ask Kenneth Short why he and Rachel charged off down the organic path and he smiles. "It's to offset our race cars and other carbon-emitting activities." How many petrolheads have such a conscience?

While dairy industry professionals think they've inhaled too many exhaust fumes, their profitability and honesty can't be denied. Their figures are freely available on DairyNZ's case studies webpage for comparison.

They operate Shortland Farms, two properties in partnership with Rachel's parents near Opunake, Taranaki - a home block of 168ha with contract milkers milking 350 cows, plus grazing all young stock from both farms, and 13km away sharemilkers manage a 68ha block with 200 cows, all Jerseys. Expected herd production is 320kgMS/cow over a 280-day lactation, milking once-a-day from Christmas at the home farm and full season once-a-day at the other. All is self-contained with no imported feed brought to the properties.

They began questioning their whole operation. Why do sheep? How could the operation be simpler? In fact, why keep stock all year?
Guy, Ashley and Ian Coats

Originally, DairyNZ consultant Steve Canton suggested either maximising urea or going organic. As Rachel recalls, they were curious as they were basically organic except for sprays and fertiliser, there wasn't much to change. And there was a local community of organic dairy farmers with 20-plus years' experience to draw on. They signed up for Fonterra's organic programme in 2015 and were fortunate to attend a Fonterra Organic Study Tour to North America to check out the scene there.

The toughest part of this journey was blowback from their own industry. Rachel and Kenneth won three environment awards in 2022 and at the ceremonial dinner, Rachel's mother, the local MP, was approached by a dairy leader embarrassed she'd been placed at the same table as the weird organic dairy farmers. She explained her connection and they received an apologetic phone call the following week.

They've become great ambassadors for Halter collar technology. It's not just useful for virtual fencing, but monitors rumination, grazing, rest and movement, plus it detects when animals are cycling, as well as analyses reproductive performance. They've found many efficiency benefits for their herd. For example:

- Pre-set breaks for cows up to a week in advance. Young stock don't have collars so are still managed with normal fences.
- Pausing collars for cows about to calve so they can access the freshest feed.
- Grazing management is simple and accurate through the use of virtual fencing.
- No more tail painting during mating. Halter will detect a cow on heat and their flashing collar when they come to the cowshed makes drafting easy.

As anyone knows, once a cow jumps a break, it's almost impossible to bring them back immediately. The fencing app gives cows time to munch ahead of the herd before guiding them back – although it's very rare for cows to come out of the virtual break.

Training cows to Halter is easy, training humans to leave gates open is more challenging when minds are so trained to shut the gate. However, there's a potential problem. They heard a story about a farmer at the pub showing his mate Halter and accidently sending his cows to the shed at midnight. There's three minutes between sending instructions and cows on the move. No safety catch…

An interesting outcome is how animal behaviour changes. Cows no longer associate humans with food so don't get up when they hear approaching vehicles. When recent severe weather passed through, cows got up on cue and walked head first into driving rain, very unusual for their herd.

With a projected fee of $0.46/kgMS ($14.50/hd/mth), Halter costs around 10% of FWE, only beaten by repairs and maintenance ($0.51) and labour ($1.76), with fertiliser $0.21. That'll generate total FWEs of $4.79/kgMS for the 2025/26 season compared to last season's $4.62.

As well as multi-species crops, they've become fans of deferred grazing allowing up to 60 days recovery with multi-species pastures, whereas normally they don't dip below 30 days. No turnips, maize or silage are used and all hay cut is on the property. While other farmers panic about losing quality, they are less concerned as they can put that into another class of livestock or save it for drying off. For a week before drying off, cows fill up on bulk, rank feed as the high fibre diet keeps animals content.

> Animals bloom on grass all year round benefiting from high covers, balage and hay, all home-grown, no non-grass supplements. Each mob is kept together for better social cohesion and less fighting – Jamie and Nicky Gaddum and Robert Hasselsteiner

To renew pastures, they use an aerator and then direct drill a 20-plus species mix into existing pasture. Their mindset has changed from traditional ryegrass and clover, believing if grass went to seed they were losing quality, to enjoying seeing pastures seed and regenerate. Cows graze new pastures for a couple of hours, then the paddocks are left to recover for 60 days to be grazed again, then incorporated into the grazing round.

A Tow and Fert applies fish and seaweed, humates and trace minerals and they have started adding seed when applying to paddocks that require more diversity. They've found their mix of diverse pastures and soil conditioners has essentially eliminated any pasture pests. Pastures are deeper rooting resulting in less ponding during rains.

The Shorts believe their greatest insights for dairy production came from learning soil function. Improving soil generates healthier animals, meaning they're not worrying about treating animals for problems.

Now they're considering permaculture with eco-design consultants Greenbridge developing landscape plans for shelter belts to reduce wind flow. These hedgerows are also for browsing.

Despite being ignored by industry professionals, the Shorts demonstrate resilience going down regen and organic paths with costs less impacted by inflation. After fertiliser prices doubled in the 2023/24 season, their FWE came to $3.68/kgMS, the lowest of all DairyNZ case studies for that season, resulting in a profit of $5031/ha, the second-highest.

Eco-profit Observations

- Figures for their organic dairy farm are freely available on DairyNZ's case studies webpage for comparison

- Challenged to maximise urea or go organic, they were curious as they were basically organic except for sprays and fertiliser.
- The toughest part of this journey was blowback from their own industry.
- They've found many efficiency benefits for their herd using electronic collars from Halter. However, there are short coming sending messages. No safety catch. Does change animal behaviour.
- Become fans of deferred grazing as less worried about losing pasture quality. Only supplement is hay, no crops or silage.
- Developed a completely different practice for pasture renewal. They direct drill into established pasture and graze at 60 days.
- Use Tow and Fert to add nutrients and seed.
- Diverse pasture eliminate pasture pests and deeper roots result in less ponding after rain. Improving soil creates healthier livestock.
- Developing hedgerows to reduce wind and create browsing opportunities.
- Going organic minimises risk to input inflation.

25. GMO Ryegrass only Treats Symptoms

The scientific community has changed our life more in this century than any parliament, and yet it feels obliged to justify nothing.
John Ralston Saul. (1992). Voltaire's Bastards.

Whenever people ask about GMO ryegrass, I remind them of a statement credited to retired AgResearch scientist Tom Fraser. Upon addressing a farm group talking new clover cultivars he remarked "You already have about 100kg/ha of huia (white clover). Do you think sowing 3kg/ha of a new clover cultivar will make a difference? How will you tell them apart when they germinate?"

My question is how would it be any different when introducing GMO ryegrass? With up to 60,000 seeds per square metre, many new seedlings will be ryegrasses from old cultivars. Sowing rates are usually around 2,000 seeds per square metre and as research shows, similar production can be obtained whether there's 40 or 400 pasture plants per square metre.

Despite all the talk of precision agriculture, no fertiliser or agrichemical company guarantees their products stay where put – Valuing Rainfall

As almost everyone following this space knows, the majority of GMO plants are designed to be herbicide-resistant. The only way to maximise projected profits is to continually spray out all other pasture species. Twenty-five years of herbiciding GMO crops in North America resulted in over 160 herbicide-resistant weeds. Just how this outcome helps farmers is never mentioned.

This is up there with feeding cows mashed brains and spinal cords - science where only farmers are held accountable. Does GMO ryegrass address the root causes farmers face?

Overseas research is continuing to challenge monocultures of any pasture species; that pastures and crops produce more biomass when planted in polycultures than on their own. Different plants with diverse leaf shapes, angles and volumes result in pastures photosynthesising more sunlight, producing more and different kinds of exudates to enhance a living soil community that builds better water-holding capacity and fertility.

New Zealand research also shows livestock production improves with multi-species pastures, particularly deep-rooting plants. There is no mention of whether the roots of these ryegrass plants go deeper or are more vigorous in growth than non-GMO cultivars.

Methane? All grass whether grazed or not releases methane upon its breakdown. Methane is a short-term GHG. Retired Australia CSIRO scientist Walter Jehne is one of the best communicators explaining how grasses reduce methane. Transpiring grasses release water vapour loaded with hydroxyl ions. These ions convert methane into carbon dioxide and water. Climatologists have known for more than 60 years water vapour accounts for 95% of global heat dynamics and that carbon dioxide is 4%. The reason climate scientists can't see the elephant in the rumen is that it's too big to fit.

So what do farmers really need to know about methane? Transpiring grasses release 100 times more hydroxyl ions to deal with methane from livestock grazing than that exact same pasture can produce. As Jehne points out, to save the world from a Siberian methane burp we need more grasslands grazed by livestock. Though this is acknowledged by the IPCC, it's never been explored.

Drought? On-farm most droughts arise from poor decision-making around grazing practice. Just how this new ryegrass will prevent farmers overgrazing it is unclear. If farmers continue existing practices that exacerbate drought conditions this ryegrass will not correct their habits.

Research and consultants constantly mention it's not types of technology, crops, livestock or whatever else farmers use which determines profitability, its financial discipline as well as the ability to observe surroundings and act which matters.

So how do we minimise risk in New Zealand? The simplest way is to run sheep and beef cattle as they have almost opposing market cycles – Market Cycles and Price Swings.

More sugar? Higher sugar levels result in plants maturing faster and livestock on shorter rotations. What will that do for soil compaction, reducing irrigation and soil aeration, or the constant need for pasture renewal? Shorter rotations reduce root integrity, resulting in livestock pulling plants out of the ground. GMO ryegrass grows faster but there is no mention of whether mean annual production also increases, higher or lower levels of mineral nutrients are required or speeding up plant metabolism increases pasture longevity.

Even sugar ryegrasses change their composition when excess nitrogen fertiliser is applied by reducing stored complex sugars required for fast regrowth after grazing. It's this energy reserve that puts up the first flag leaf. Without it recovery periods between grazing lengthen.

There also no mention of whether livestock prefer this grass over others. Again in North America, wildlife refuse to graze GMO crops because they are bitter and maize cobs contain chemicals like formaldehyde. In feedlots, farmers notice livestock behaviour changes once eating GMO feeds. They become restless and agitated. Furthermore, death rates escalate when livestock are confined to GMO feed for more than 75 days.

Farmers here report similar outcomes grazing fodder beet and swedes sprayed with chemical cocktails. Just like in North America, no research is ever done on how these chemical cocktails accumulate poisons over time and their impacts on livestock health and performance.

Does GMO have huge potential? As Australian colleague Graeme Hand argues, imagine if science could produce a grain that thrives in a perennial grassland, doesn't require additional fertiliser, pesticides, fungicides and herbicides, with straw that properly decomposes on soil surface contact, seeds remain true for the following season and people get healthy eating the grain and suffer no unintended consequences. If that ever happened, it would significantly change the nature of agricultural services.

But the reason this is highly unlikely is because of the difference between how nature works and how humans think. Luckily, we have farmers now looking for solutions to bypass this risky and expensive technology.

Eco-profit Observations

- How would farmers know the difference between a GMO grass and one germinating from the resident seed bank?
- Most GMO crops in North America require herbicides to be effective, resulting in over 160 herbicide-resistant weeds. How does that help the farmer?
- Diverse pastures capture most sunlight and produce a greater array of exudates to support soil life.
- Livestock production improves with multi-species pastures, particularly deep-rooting plants. There is no evidence GMO ryegrass improves this trait.
- All grass whether grazed or not releases methane upon its breakdown.

- Transpiring grasses release hydroxyl ions which break down methane into water vapour and CO_2.
- If farmers continue their existing practices how this new ryegrass will prevent farmers overgrazing it is unclear.
- Higher sugar levels result in plants maturing faster and livestock on shorter rotations. What will that do for soil compaction, reducing irrigation and soil aeration?
- North America wildlife refuse to graze GMO crops because they are bitter and maize cobs contain chemicals like formaldehyde.
- No research is ever done on how these chemical cocktails accumulate poisons over time.

Adaptation: GM Ryegrass just treats Symptoms. Farmers Weekly, August 13, 2018, p29
https://issuu.com/farmersweeklynz/docs/farmers_weekly_nz_august_13_2018/31

26. Jeremy Casey and Kim Solly - Science and Sustainability

While running 2,000 cows near Winchmore, Canterbury, Jeremy Casey and Kim Solly were having a lot of animal health issues. They believed it was due to what the cows were eating, so ventured into minerals and alternative fertilisers such as Eco-N and Donaghys LessN. This was driven by their own curiosity and from mixing with Lincoln University staff at field days.

When the owner's son came home, the sharemilking contract ended. As Kim and Jeremy started looking for new challenges two farms, originally one farm, came up for sale either side of Back Track Road, Methven, and they bought both with an investor. Wairoa, the northern farm (210ha), interested them first but it needed scale to access an irrigation right. Whakapono, the southern farm, was smaller (155ha) so that became a testing site for alternatives to reduce risk.

Around this time they were introduced to Neal Kinsey and the Albrecht cation balancing approach to soil management by Kim's father, Merv Solly, through his involvement with Golden Bay dolomite. By now they were working with Top Soils consultant Don Hart, a local cropping farmer and Albrecht consultant. Top Soils organised fertilisers for Whakapono while fertiliser company Ballance oversaw Wairoa. With connections to Lincoln University, they got funding to measure environmental, production and financial differences between more biological-friendly and conventional fertility approaches.

Funding came from Sustainable Farming Fund, DairyNZ and AGMARDT and totalled more than $300,000 for a three-year trial. They set up a steering committee and pulled in Lincoln University staff to measure and report on the findings. They included people from either end of the fertility argument spectrum.

The idea was simple: Keep everything as similar as possible to see if fertility regimes made a difference. This meant having same stocking rates (3.6hd/ha), grazing rotation lengths, amount of grain fed in milking sheds and weekly pasture walks with staff going everywhere together. There were a lot of in-depth plant, soil and ecology measurements. They held 4-5 field days every year.

The biological farm grew as much grass with less applied nitrogen because of more clover. However, criticism emerged for being low nitrogen users and therefore not a typical Canterbury irrigated dairy. That said, there was lower leaching and less GHG emissions, so it was more efficient. The biological side received additional fertiliser with more calcium, magnesium and sulphur while getting slightly less phosphorous and nitrogen. That lifted costs.

Jeff learned that the half-life utilisation of elements is determined with conventional fertilisers, not biological ones or combinations. When formulas are changed, so does utilisation and efficiency – Jeff Broomfield.

It also improved animal health with empty rates down 2% and fewer deaths, but it had issues with magnesium in spring. To combat that they ran a normal spring pre-calving regime but dressed pastures with dolomite post-calving. Putting magnesium in water troughs didn't address milk fever or ryegrass stagger issues.

For the 2014/15 season comparison, biological vs conventional production per cow was 499kgMS/hd vs 492kgMS/hd. Better animal performance coincided with lower pasture crude protein and fibre but higher dry matter, sugars and ME. The biological side produced healthier grass.

They were able to run the trial for another two seasons, gaining data over five years, but no huge difference emerged between the farms. No statistical difference arose regarding pasture production, although ecological communities were more numerous and stronger on the biological farm. It also had less blocky soil structure, better animal health and the staff preferred working there.

Jeremy observed that the biological side running out of puff as it slid further away from readily soluble fertiliser. It coincided with slow clover production during shoulder seasons. Yet, one thing that both properties never got to lift was soil organic matter (SOM). Having around 3% meant there was a buffer of 60-90kg of nitrogen to drive spring and summer growth. This perpetuates fertiliser use in conventional irrigated systems. Even at 5% SOM, nitrogen storage would increase to 100-120kg/ha, lifting production and reducing costs.

Furthermore, mirroring conventional grazing management did little to push more carbon into the biological farm's soil and build capacity. Did removing opportunities to run longer rounds reduce potential to stir a significant difference in other ecosystem services? Despite pressures to change, they decided to focus on whether minerals and soil fertility approaches made the difference, rather than other farm management practices.

Clover was a huge difference between the farms, reflecting the type of nitrogen fertiliser used. This prevented waves of growth caused by urea applications and reduced many of the problems that could bring. This meant less stress. They did learn to apply potassium over three applications to stop pasture hardening off. Potassium delays pasture senescence.

The project's final report concluded the biological farm produced more positive animal health and production outcomes, thereby holding great potential over conventional programmes. However, it didn't generate enough cash during the research period to pay for its fertility programme and so had reduced attractiveness to sceptical dairy farmers.

One presentation reported that when both fertility systems reached their ideal chemistry, profitability was the same. This statement is insightful because it confirms no financial penalty and creates no drive to change to healthier, kinder systems. It maintains the status quo, with no interest from banks or policy makers. It also reflects background mumbling regarding research projects.

> Reduced labour, animal health and supplements; no PKE, grain or maize produced a $30,000 cash surplus that couldn't be accounted for by management agency Farm Right – Roger Slee.

Was this project a victim of systemic academic bias? Focusing on one element alone (mineral fertiliser) did nothing to separate both properties in farmers' minds. When farmers experiment, they change several things to accentuate results because it's outcomes that matter. Did the structural methodology focused on single factor analysis railroad its outcome?

Jeremy and Kim are proud of their farming achievements and of developing meaningful relationships with scientists. Their momentum challenges old beliefs with new conversations as they continue to find kinder solutions to Canterbury irrigated dairy.

Eco-profit Observations

- Animal health problems drove curiosity into alternative solutions.
- Bought two farms with intention of trialling ideas on one to compare with the other. Got industry backing for a trial involving people from either end of fertiliser management.
- The study kept everything the same except mineral fertiliser applications.
- The biological farm grew more pasture, had lower leaching and GHG emissions, lower empty rates, fewer deaths, lower crude protein, better animal performance, healthier pasture, better soil function and better ecological communities. Staff preferred working there.
- The biological side had higher fertiliser costs.
- No statistical production differences over five years.
- Low natural nitrogen levels associated with low soil carbon levels and coupled with slow clover growth was reducing biological farm production.
- Inability to change grazing regime to stimulate carbon sequestration and other ecosystem services a possible design fault?
- Clover nitrogen reduces boom/bust experienced with artificial applications.
- Potassium applications reduced pasture maturity and lengthened growing season.
- The biological farm didn't generate enough cash to pay for its fertility programme and this reduced its attractiveness to dairy farmers.
- The conclusion that when both fertility systems reached their ideal chemistry profitability was the same confirms no financial penalty and creates no drive to change to healthier, kinder systems.
- Did the structural methodology to ensure single factor analysis railroad its outcome?

27. Peter and Mandy Paterson - Challenging Industry Bias

A way to success is to surround yourself with a community of supporters. Organic dairy farmers Peter and Mandy Paterson have hosted many organic professionals to keep the momentum going. This non-nonsense couple tenaciously demonstrate what's possible and what's myth in an industry biased against organic production.

Peter and Mandy milk 190 cows on 80ha near Patetonga in north Waikato. A family health scare got them questioning agrichemical use in food production. They started converting to organics in 2015 after a 15-year journey with biological methods. They were told going organic would drop production by 30-40%. It's never happened.

Tony calculated that area (15% of their entire operation) had to generate more cash than from the original property just to cover interest payments – Tony and Angela

Mandy regularly challenged DairyNZ extension agents at field days on achieving higher production without urea. Finally, a DairyNZ facilitator, who had no time for organic production, organised a field day for Mandy to demonstrate what many thought impossible.

More than 90 people arrived for a day during drought in March 2018. Their farm was the only green patch for miles. Many local properties following industry advice were already feeding out. Guests walked into pasture standing around 3,000kgDM and still growing. Autumn calving was ending and their system proved they were viable when many other operations were not.

The facilitator was so impressed he asked other DairyNZ staff to visit. None did, so in frustration he quit and went back to being an accountant. A year later, Fonterra, which collaborates closely with DairyNZ, catered for a North American tour group from Organic Valley to spend an evening with the Patersons. It's interesting how Fonterra's attitudes to organic production contradicts the national dairy extension agency.

What's their secret to success? Peter and Mandy fully feed cows. Visit their farm in winter and you'll find cows consuming 24kgDM, 14kg in the paddock and the rest in their herd home. With a stocking rate of 2.2 cows/ha, they easily fully feed animals. By some estimates, 80% of New Zealand dairy cows are not fully fed. Cows peak around 2.0kgMS/cow in autumn after calving in March, ticking down to 1.8 through winter and into Christmas.

Mandy says the herd home is what creates flexibility to promote pasture growth. The easiest way to grow more grass is to keep cows off paddocks longer, understandable when over the boundary is a 25,000ha swamp. Cows are fed in the herd home for 10 months. All supplements are fed there with only maize grain fed in the milking shed. There is no wastage, everything is consumed.

They make most supplements on-farm and while it's easier to buy supplements now than 15 years ago, it's expensive. They can have up to 600 bales of silage as backup. Winter milk contract premiums largely pay for their supplementary feed.

Many experts said they would never grow an organic maize crop, but they do it easily, 21-23 tonne this year, 2024. Last year much of the farm was under water resulting in half the maize crop they normally expect, 100 tonne. Now other locals plant maize using inter-row weeders to make it work, a task they will do six or seven times

An important observation is cows start eating pasture once inside gateways instead of bouncing off paddock fences looking for something. Pastures contain up to 16 species, including timothy, red and white clover, sheep's burnet, chicory, oats, cocksfoot, prairie grass, tetraploid and diploid ryegrasses. The mix includes 20kg oats with 20kg pasture mix and sown without slug bait or other pest controls.

He found synthetic fertiliser significantly impacted soil biology and health by removing the abilities of fungi and bacteria to function and build organic matter. It also tied up various trace elements, causing animal health problems and hindering performance – Andy Bunny.

Oats grow up first and perennials grow underneath. Cows graze oats four times for perennials to emerge and never get bloat. Pasture nitrate colour kits confirm if pastures are safe for grazing.

The big wet of 2023 stripped soils of any soluble nutrients. Peter Norwood, their agronomist, makes recommendations based on Albrecht tests and hair analysis. One animal has been hair tested for 12 years. From there, Mandy finds out what's available. The wet stripped all sulphur off the paddocks, so in spring they applied a mix of kelp, molasses, fish hydrolysate, magnesium sulphate, copper sulphate, zinc sulphate and cobalt sulphate, BD Max 501 and Boosta Bio.

They only need 40 heifers for replacements but have experienced fertility problems with up to 25% empties, though they are now closer to 10-15%. Mating with Halter collars achieved 90% conception in the first round by picking up an additional two cows a day while identifying non-cyclers early. Mating starts on May 25 with sex semen for a week, then normal semen is used till July 12, with natural mating until August 1.

Calving with Halter is easy. They turn off collars for calving cows to move into a fresh break. If they strike a stormy night, they can move cows to another break, all from the comfort of their kitchen. In fact, most shifts are executed by grandson Thomas, who is studying at Lincoln University, 1,000kms away.

The 2023 big wet created animal health problems. Cows presented smelly mastitis and blood tests confirmed ecoli. Nothing, including Engemycin and Medicam, worked here. Modern drugs just masked symptoms because once removed animals were sick within hours, and then died. Industry cures became an addictive expense.

They find True Health remedies and homeopathy works for 90% of animal health cases. While combining both with organic hay tea settled sick stock, they still died in this instance. To make hay tea, they'd half-fill a drum with warm water, add a section of hay from a small bale, then leave to seep all afternoon for animals to drink in the evening. Then do again for the morning.

Peter and Mandy have built an organic system defying industry expectations and agrichemicals. It's an example of how commitment and self-belief creates a production system for others to follow.

Eco-profit Observations

- They were told going organic would drop production by 30-40%. It's never happened.
- A DairyNZ facilitator with no time for organic production, organised a field day. Over 90 people arrived during drought. Theirs was the only green farm for miles.

- The facilitator asked other DairyNZ staff to visit. None did, so in frustration he quit.
- They easily fully feed animals whereas some estimate 80% of New Zealand dairy cows are not.
- The herd home is what creates flexibility to promote pasture growth.
- Winter milk contract premiums largely pays for their supplementary feed.
- They easily grow a 21-23 tonne organic maize crop.
- An important observation is cows start eating pasture once inside gateways instead of bouncing off paddock fences looking for something to eat.
- Oats grow up first and perennials grow underneath. Cows graze oats four times for perennials to emerge and never get bloat.
- Their agronomist makes recommendations based on Albrecht tests and hair analysis.
- Mating with Halter collars achieved 90% conception in the first round by picking up an additional two cows a day.
- With Halter, most shifts are executed by grandson Thomas who is studying at Lincoln University 1000km away.
- They find True Health remedies and homeopathy works for 90% of animal health cases. Hay tea is better than modern drugs for settling sick stock.

28. Karl Steiner - Isolation and Innovation

Karl Steiner's 140ha dairy farm near Galatea, Bay of Plenty, is a hotbed of ideas and experimentation in a challenging location. His curiosity might be hereditary as he's a descendant of biodynamic farming pioneer, Rudoff Steiner. He turned to organics because so many commercial fertiliser products didn't work on his pastures.

His herd of 180 mostly Friesian cows produces 70,000kgMS (390kgMS/cow) on a System 1 set-up where he raises his own replacements. Pumice soils handle 200mm rainfall events easily without puddles, making it a great wintering property, though pastures dry out in summer. There's no irrigation but a couple of other ideas allow him to handle the summer dry.

One is pushing summer rounds out to 60 days. Meeting Kiwi Fertiliser's Brett Petersen 12 years ago and using their carbon blend achieved this within three years. Longer rounds result in cows consuming a higher fibre diet which reduces many animal health problems.

Another is pasture mixes. For example, he broadcast silver beet and daikon on pasture with fertiliser. Neither fully expressed themselves as they need 120 days, so he threw the surplus seed into tree rows. All came up and outcompeted grasses there. Cows now graze them through electric fences.

But it's his eclectic ideas around energy which makes a visit interesting. One project is making hydrogen. It requires electrolysis, involving positive and negative plates in water, to generate hydrogen with caustic soda to make bubbles flow better for gas collection. This process requires a compressor to store gas in cylinders capable of withstanding 10,000psi.

However, New Zealand's strict rules on compressing hydrogen gas necessitates certified systems for use in cars, trucks and tractors. While costing 20 cents/kg to generate, it requires three-phase power and multi-stage pumps to prevent hydrogen heating under pressure between cycles as it's explosive at over 220C. That means

> In looking to reduce fermented supplements he found hay making and drying technologies in Europe. He calculates that the latest hay-making regime using smaller tractors lowers contractor costs by 60% – Adrian Frei.

diesel motors require replacing injector pumps with common rail injection and a throttle body.

Karl knows how dangerous hydrogen can be. He learnt the hard way not to weld next to lead batteries when charging as that's when they produce hydrogen gas. After a flash and a bang, he ended up with acid everywhere and had to waterblast the entire workshop and its contents as touching any item burned his skin.

Eventually, he'll run everything on hydrogen but currently regulations are too confusing. Insurance companies won't cover what they can't define. Once that's done, he'll start converting a firewood-processing machine first.

His interest in clean technologies triggered an enterprise opportunity. After planting 270 fruit trees on part of the farm, an entrepreneur approached him about setting up a semi-retirement community (village is a rateable term) for rural people to stay in the area. The community has 16 tiny house sites, all off-grid on 2ha. It's one of 12 sites around the country; all located on organic farms.

The $36m^2$ portable houses are one or two room structures. All have heat pumps and composting toilets with electrics driven by solar panels. Power is stored in sand batteries under concrete floors within a stainless-steel tray, no lithium. Each farm site has a communal kitchen and laundry. One interesting technology is this building has a triple glazed window that generates bubbles so curtains aren't needed. The same window also heats water.

Tenants buy their houses for $300,000 and pay the farmer $30 a week rental to cover rates. While tenants can take their houses to other sites, it's the farmer's job to encourage people to stay.

If you think Karl mixes with some interesting people, it doesn't stop there. Through contacts he learned of the Ministry of Forestry's 40-year-old research proving land only needs 15 leafy trees/ha at 10 years growth to be carbon neutral for a dairy farm's lifetime. This is crucial to understand national policy involving carbon taxes.

With encouragement from Bill Quinn's organic farming group, he submitted an idea to the national climate policy committee in 2020, two weeks after its closing date. Then politicians excitedly called him while reading submissions; his was the only one presenting a plan for a national roll-out.

The idea came from Scandinavia. It has a simple rule of thumb: When cutting down a tree, plant another. It doesn't matter what kind of tree but a minimum of 15 leafy trees/ha per land title is required. To give an idea of scale, a farmer could cultivate a paddock 50 times per year with a 100hp tractor and still be carbon neutral. Even shopping malls in Scandinavia do this.

Unfortunately, this idea was dumped for carbon taxes.

Back on the farm, Karl uses another form of energy for animal health. It draws on electromagnetic colour frequencies similar to homeopathy's subtle energies. It is housed in PVC tubes with a copper coil on the lane running parallel to his herringbone milking shed and influences up to 350 metres away. It's called colour therapy and he's used it for 26 years to treat cow infections.

> By lunchtime he'd cancelled an order for a new plough and ordered a direct drill. Then a Soil Foodweb course encouraged him to buy his own microscope – James Halford

The result is no teat spray is needed and no dry cow or teat seal used despite a somatic cell count of 250,000-300,000. Cows have no problems drying off suddenly. In addition, Karl believes environment doesn't cause mastitis, but bacteria from human skin. He's heard smearing forearms with calf saliva instead of iodine or rubber gloves neutralises bacteria. After all, the only natural fluid touching cow teats is calf saliva.

If isolated farming communities breed independent thinkers and doers, Karl is a shining example of what is possible.

Eco-profit Observations

- Turned to organics because so many commercial fertiliser products didn't work on his pastures.
- Longer rounds result in cows consuming a higher fibre diet which reduces many animal health problems.
- Uses tree rows to grow pasture species that require longer recoveries than can be achieved in paddocks.
- Hydrogen costs 20 cents/kg to generate but requires expensive equipment and regulations are currently foggy.
- Set up a semi-retirement community using off-grid tiny homes as part of a nationwide network.

- Ministry of Forestry's 40-year-old research proves land only needs 15 leafy trees/ha at 10 years growth to be carbon neutral for a dairy farm's lifetime.
- Yet the tree planting policy idea was dumped by national climate policy committee in favour of carbon taxes.
- Has successfully used colour therapy for 26 years.
- It's possible smearing forearms with calf saliva instead of iodine or wearing rubber gloves neutralises mastitis bacteria.

29. Louise Coats - Cause or Cure

North Waikato dairy vet Louise Coats recalls the moment she first began to question the reasons for health issues. Driving early afternoon, a tree suddenly appeared in front of her. This car crash set her on a trail of discovery, the first step was finding that high carbohydrate lunches make O blood group people very sleepy.

That led to researching human, plant and soil nutrition, chasing experts such as Elaine Ingham and Arden Anderson, discovering Albrecht cation balancing, then biodynamics and then for 25 years biological farming. To many of her farming clients she wears a farmer's hat instead of a pharmaceutical one. Often, she's looking for causes rather than cures because it's cheaper for the farmer.

> When her meticulous research delivered results favourable to product promotion, it was publicised. Anything not was silenced – Angela Payne.

During her veterinary training she had little exposure to nutrition, let alone any appreciation of what nutrition can do for improving livestock health. Over many years Louise noted anecdotally that trees may have anthelmintic properties. There are certain periods of the year when some bark types are very attractive for cattle due to fibre, minerals or photonutrients. Some of her clients swear by privet seedlings, red mapou and flax.

Upon meeting lifelong partner Greg and checking out his dairy farm storeroom, she was surprised at the shelves of metabolites for treating milk fever and grass staggers. Over the years, the family herd has been used to explore various treatment or preventative programmes.

For example, her interest in Ivermectin came from cutting cancers from cows' third eye lids. Once drying cows off and administering a flukicide (which contains Ivermectin) along with an additional Ivermectin treatment a month later, she came back to inspect her work to find they'd healed completely. The standard Ivermectin dose is 1ml per 50kg lwt. Ivermectin stimulates immune systems so it's more than just a parasitic treatment.

Then in 2013, her family's herd got neospora. She traced the most likely cause of the cow abortions to duck shooters' dog dung deposited on a canal bank.

Over 90% of calves born from infected dams will carry neospora, perpetuating its cycle. Up to 40 of the 150 animals were carriers so in the ideal world their heifer calves should not be reared as replacements. Heifers carrying neospora infections would start aborting from February right through to calving time.

Adding 5ml per head with calf milk at each feeding resulted in firmer stools. The smell of ammonia disappearing completely from the calf shed within 36 hours – Andrew Tabener.

In 2023, Louise conducted an experiment involving all neospora-carrying heifer calves, treating them with an Ivermectin injection and then another before they joined the herd. All 13 took their calves through to full-term without issue. It's a pleasing result and Louise will replicate this experiment again but it requires a large-scale trial for statistical significance.

Other problems Louise sees involve copper, selenium and B12 deficiencies in all classes of stock in her district. Animals with low liver stores of copper often have front leg fractures from September onwards. Heifers walking to cowsheds can come in with a broken leg despite never tripping or stumbling, especially at peak lactation. Louise wonders if agrichemicals like glyphosate are responsible for reducing mineral availability to livestock.

Retired emeritus professor and overseer of Monsanto's programme at Purdue University for 27 years, Professor Don Huber, toured New Zealand in 2015 voicing his worry about glyphosate tying up elements of life like copper, manganese and zinc, effectively shutting down metabolic pathways. These three elements are vital for many enzyme reactions. For example, animals with insufficient copper will not efficiently convert grass into body condition. Many cows going down with milk fever are often copper-deficient.

Louise fears other animal health problems lurk for cattle drinking from the Hauraki Plains town water supply. Unusually large numbers of dairy cattle drink water from this public scheme, which is soon to be medicated with fluoride. Fluoride prevents proper construction of bone lattice when cows go through periods of depositing and absorbing layers of calcium phosphate due to intense milking. She fears that drinking more than 100 litres of water daily, they'll take in enough fluoride to increase susceptibility to leg fractures. The stock water is also treated with chlorine, which reduces iodine availability.

Another debilitating disease is Johnes. In New Zealand, Louise observes it's hidden on farms as farmers know the symptoms and cull those cows. Without a screening programme farmers find they're culling an ever increasing number of animals due to recurring diarrhoea and weight loss. Johnes cases increase during times of stress such as calving or peak lactation. Infected cows often infect their calves from colostrum and calves up to six months become infected from cows shedding bacteria in dung.

Testing milk for Johnes during herd tests identifies three categories: High positive, positive and suspicious. Vets found blood tests on high positive cows correlate with a milk test. This allows farmers to sell cows before they lose condition.

Prior to milk screening being available, Louise treated Johnes by drying cows off, treating with a 36 gram copper bullet orally, along with one packet of Parnell diarrhoea powder, and then with another diarrhoea powder a day later. Treated animals then graze rough mature pastures away from paddocks young stock have grazed. These measures stop diarrhoea and regain body condition before culling.

Unlike many in her industry, Louise developed an interest in homeopathy when the founder of Homeopathy Farm Support, Tineke Verkade, suggested she try using just one remedy, arnica. She found clients were open to trying alternative treatments. Over time, clients report back that livestock appear to recover faster after surgery.

After a lifetime of veterinary practice, you can't accuse Louise of falling asleep at the wheel.

Eco-profit Observations

- Noted anecdotally that trees may have anthelmintic properties and why animals eat bark at certain times of year.
- Ivermectin stimulates immune systems so not just a parasite remedy. Removed eye lid cancers didn't reappear after Ivermectin cure.
- Ivermectin appears to prevent abortion in neospora-infected heifers.
- Wonders if agrichemicals prevent uptake of elements such as copper, selenium, cobalt, manganese and zinc.
- Fluorided town supply water has potential to disrupt bone metabolism for dairy cows. Chlorine disrupts absorption of iodine.
- There's a relationship between testing for Johnes in milk and blood to help farmers identify which animals to cull.
- Livestock appear to recover faster from surgery under homeopathy.

30. CHALLENGING FARMING CULTURE

Copernicus, Galileo, and Kepler did not solve an old problem, they asked a new question, and in doing so they changed the whole basis on which the old questions had been framed.
Sir Ken Robinson.

There's plenty of confusion over regenerating (I prefer a verb to an adjective) farming because although we can describe it, it remains undefined. This frustrates agricultural science because all production is accounted for in recipes, whether industrial or organic. So why is it difficult to define? Let's use cooking as an example.

Combining sugar, eggs, flour and milk together creates a cake mix. Mixtures can be ready-made, foolproof and substitute natural ingredients, reflecting the industrial model. Whip the elements together, pop it in an oven and 40 minutes later a cake emerges.

But organic devotees use raw not white sugar, pastured instead of caged eggs, spelt instead of wheat and raw milk to make their product. Bakers might end up using less of these ingredients to create delights, but they're more costly to buy. Results can vary because these ingredients do not always behave as consistently as standard ingredients. This creates risk.

So where does regenerating fit in? Regenerating challenges both by changing recipes and ingredients. It uses any combination of the above because it's the result that matters - the cake's form, texture and taste. To achieve that means tinkering to bake the best cake, whereas industrial and organic production rigidly adhere to inputs and methods to bake a standard cake, no matter the circumstances. This is the backbone of industrial advice.

At the core of regenerating is not the method; cake ingredients, mixing, baking and tools, nor how these elements are combined. What regenerating does is help farmers understand what creates outcomes instead of blindly accepting what they're told. It's the Naked Chef approach to agriculture.

Everybody remembers Jamie Oliver bursting onto TV 25 years ago. Young, approachable and a fresh way of relating to food that inspired many tired of masterclass cooking shows. His eagerness to sip sensations, inhale aromas and feel textures by using his hands to mix, squeeze and get dirty with food changed how audiences saw themselves.

He embodied how families act in kitchens, using commonly found ingredients and principles to create repeatability, cheaply. Gone were gourmet continental classics and spotless stovetops. Cooking didn't require Michelin stars, detailed recipes, expensive ingredients and hours in the kitchen. Jamie encouraged viewers to engage with the creative process by licking their fingers and savouring changing flavours while adding ingredients. Why stress over perfection when the fun is in keeping things simple?

Regenerating agriculture celebrates this same connection. Farmers engage curiosity and intimacy by using local, cheap resources (nature) and principles. In taking back control of doing things in paddocks, it encourages intention; farmers creating outcomes rather than being passive observers of ready-made techniques and products. Linking intent to values creates a brand.

> When technologies failed, Tony and Jerrie Tipton were the last call and yet every time they completed a project their work astounded everyone – Livestock Save the Planet.

So what has been critiqued about Jamie's intent? He's been condemned for over-simplifying cooking and having audiences believe it's more cost-effective than it is. And in pushing low-cost ahead of access to time, skills, tools and pragmatism, he's alienated some target audiences. Yet his vision of food as health remains a popular catch-cry because that's its purpose, not entertainment.

Industrial ag criticises the term regenerating with similar arguments stating it's not as profitable or as cheap as promoted due to lack of convenience and consistency compared to industrial methods. However, industrial ag's focus is production output, not relationships between human nutrition, soil health and profit. Furthermore, New Zealand industrial ag states farmers are already regenerating but there is a missing attitude which Jamie's show demonstrates in spades, connecting to wholesome outcomes.

Regenerating attracts farmers burned by industry information and policy. Many are curious to create different outcomes for their operations rather than options delivered by a monoculture of automated advice and professional packages. They are taking control because it's impossible to get accountability and apologies from PhDs.

The No 8 wire mentality comes into play, as with so many aspects of farming. Farmers carry spades to dig holes and check soils for themselves instead of accepting reports on paper. They brew biological soups, sow crop cocktails and boost bird and bee numbers. Soil and seeds slide between their fingers without fear of synthetic poisons. They see themselves not only as producers but custodians of their land and communities.

> Any dairy farmer grazing "silage grass" with a pre-grazing residue of 3500kgDM/ha or even higher knows, production doesn't nose dive, as professionals believe – Consultants Breed Costs.

Industrial ag bites back for the same reasons McDonalds doesn't use Jamie's recipes. Regenerating farming requires nuance and self-determination that staff climbing corporate ladders are unable to execute in teams blinded to anything but production bonuses. Because industrial ag relies on efficiencies for turnover to excel with very thin margins; there's little place and desire to create from scratch. Recipes are easier and simpler for training and determining employment and pay. Paper and professionals dominate. Its why efficiency is so expensive when it fails.

Owner operators have more flexibility to substitute profit for experience but require wider margins to value resilience over revenue. Therefore, these farmers look to lower costs by employing nature instead of fighting it. This requires learning new skills to observe and confidence to act quickly. Unfortunately, farmers invest in anything but themselves.

Because regenerating blends so much from industrial and organic models, there's no hard edge for a definitive boundary. Like The Naked Chef, regenerating peels away non-essentials to create new mindsets for farmers to intentionally engage with nature. The focus goes beyond production by understanding how to benefit from ecological relationships. In filling this information void, regenerating shifts the power of learning from industry to farmers. That's why academics and policy makers remain irritated, because defining regenerating would acknowledge their own failings.

Eco-profit Observations

- Regenerating means tinkering to bake the best cake, whereas industrial and organic production rigidly adhere to inputs and methods to bake a standard cake, no matter the circumstances.
- Regenerating helps farmers understand what creates outcomes instead of blindly accepting what they're told.

- In taking back control of doing things in paddocks, it encourages intention - farmers creating outcomes rather than being passive observers of ready-made techniques and products.
- Industrial ag's focus is production output, not relationships between human nutrition, soil health and profit.
- Regenerating attracts farmers burned by industry information and policy.
- Owner operators have more flexibility to substitute profit for experience but require wider margins to value resilience over revenue.

31. Peter Berryman - Family Business Lessons

Peter Berryman recalls how to get fired instantly on Timaru fishing trawlers in the 1960s – have a morning wash. That removes nature's sunscreen, skin surface fats and oils, increasing the risk of severe sunburn. Fishing requires 100% concentration and sunburn distracts so much it could be a safety hazard. Lesson: Resilience comes first.

Venturing to the South Island from Edgecumbe, Bay of Plenty, Peter's experiences brought home many insights. One involved a crop of wheat where the Canterbury farmer allowed him to trial doubling superphosphate to 300cwt/acre on the headlands. The crop grew like crazy but at harvest produced shrivelled seed barely fit for chickens. It was only when the header left the headland that the crop yielded something worthwhile. Lesson: Less is more.

After arriving home to his family dairy farm in the early 1970s, he dropped super and potash and stuck with lime and dolomite. After two seasons of drought, he put on a tonne of lime/acre. The next season he had hay when other farmers had little.

Allows pastures to naturally seed so paddocks don't germinate weeds and need expensive and risky re-grassing – Bryce and Aneta Lupton.

There's been no superphosphate since 1980. Instead, he's targeted base saturations of 65-70% for calcium and 12-15% for magnesium, plus added a few trace elements and bacteria twice a year. He hasn't drenched for worms since the early 2000s, and has seen no grass grub for years. Lesson: Address root cause.

At a Grasslands conference at Whakatane in 1991, one man with figures was verbally attacked by other scientists. Peter walked over to him to pick his brains. In the 30 years since, he's used Burt Quinn's RPR. Lesson: Don't follow the crowd.

The RPR has produced soil very different from his neighbours. When a group of scientists visited 15 years ago, they dug holes either side of the fence. The neighbour's pasture roots went down 75mm but Peter's went 700-900mm.

Then the scientists put soil lysimeters on both sides of the fence to measure nitrogen leaching. Within two years scientists removed them as outcomes between the Berrymans' fertiliser regime and conventional were so different. Industry experts found comparing fenceline measurements embarrassing.

Differences are easy to see in the dry. The Berryman property takes much longer to brown-off than neighbouring properties. Today, digging up and putting a cube of soil in a tub of water will initiate a wicking process visibly drawing water into soil humus, demonstrating its superior ability to absorb and retain moisture. It's impossible to find a conventional property that does this. Lesson: Let nature compound services in your favour.

> It was more versatile; he could spray it on soil, put it in water troughs and drench stock. And he could do all that himself, no more contractors – Tim Hawke and Murray Weaver.

However, there's been one episode where the porous soil structure in a low spot succumbed to a rising water table due to an extremely wet and stormy 12 months. Peter believes a decline in pasture production was due to anaerobic conditions killing off pasture roots and soil biology. It took the following 12 months for those paddocks to recover their productivity.

On a 70ha milking platform, plus a further 11ha runoff for grazing and making supplements, milk production for 190 cows is more than 550kgMS/cow, according to herd testing. Balage is mostly fed out in winter as all stock stay home and during spring and autumn to lengthen pasture recoveries.

Early summer recoveries are 17-18 days as silage paddocks are still out. They seldom go over 30 days in summer. Yet they grow around 19,000kgDM/ha annually, largely through soil correction. From the 70ha milking platform, their herd produces over 100,000kgMS. Lesson: Don't judge farms by their fertiliser.

Daughter Liz and son-in-law Danny started 50/50 sharemilking in 2005. Not long after, their bank wanted them to go bigger, then encouraged them into the Sharemilker of the Year competition where they were local runners-up. The judges' feedback was to get into equity farming to grow their business. With banks looking for figures that suited their growth models, and the Berryman farm taking a different approach to sustainable farming, they did not feel confident finding other like-minded business partners. Staying biological meant they could stay at home, employ no staff and concentrate on doing what they did well while paying two generations.

They've expanded by buying lifestyle blocks – including one from Peter and mum Robin - as debt is more manageable and also, because they fall under housing loans, interest rates are lower than commercial farm land. They don't know who their bank manager is because they're not aggressive risk takers. Lesson: Run your own agenda.

Because the herd is so well fed, they don't buy any products for parasites from vets. What's interesting is vet behaviour - many drive away and never ask why these products aren't needed. An encounter with copper might prove why.

Copper toxicity is fatal but American educator Arden Anderson says calcium neutralises copper. Years ago, Peter put lime flour with copper to lift ewe intake. The dusted pasture was blue but sheep survived. In another experience, Liz put copper in calf milk but one overdosed so much it produced blue dung and couldn't stand up. She drenched with dolomite and it lived. If offering livestock free-choice copper, offer dolomite free-choice too. Lesson: Some solutions are cheap.

Just as ignoring a morning wash makes skin more resilient to sunburn, so is avoiding synthetic fertilisers, agrichemicals and intestinal drenches for better production resilience. This was strengthened by avoiding bankers and their risky demands. The Berrymans are proof many business lessons are worth investigating.

Eco-profit Observations

- Resilience comes first. A morning wash on a fishing trawler increased severe sunburn.
- Less is more. Doubling the superphosphate rate reduced wheat grain weight and size.
- Address root cause. Dropping many industry interventions coincided with fewer dairy cow problems.
- Don't follow the crowd. Picking the brain of Burt Quinn changed the form of phosphate fertiliser used.
- A fence line nitrogen leaching comparison with neighbours was stopped after two years because the outcomes embarrassed industry professionals.
- Let nature compound services in your favour. Soil humus levels are so high a cube of soil placed in a bucket wicks up water quickly.

- Anaerobic conditions due to waterlogging can take pasture production up to 12 months to recover.
- Run your own agenda. Staying biological meant they could stay at home, employ no staff and concentrate on doing what they did well while paying two generations.
- Expanded by buying lifestyle blocks which fall under housing loans so interest rates are lower than commercial farm land.
- Don't buy any products for parasites from vets.
- Some solutions are cheap. If offering livestock free-choice copper, offer dolomite free-choice too.

32. Mark, Judy and Keeley Pudney - Farming Like Old Fellas

Imagine buying a dairy farm and then having 10-15 cows going down with milk fever each day. You end the season losing 10% of the herd. What would you do next?

The vet treated the symptoms - at increasing expense - but never found the cause of the problems. Eventually, Keeley Pudney recalls, it was the discovery of a shed crammed from floor to ceiling with urea bags that gave them the answer. The farm was burnt out from too much nitrogen.

> At 18% protein on 18kgDM/day, a cow excretes up to 140 grams of nitrate. At 22-23%, that rises to 500g, at 28% it's a kilogram, over 30% it's 1.5kg with 40% squeezing out 2kg – Jeremy Talbot.

Keeley's father Mark began asking around and started applying boron and burnt lime. There was a slight improvement in herd health but more was needed. It was the start of a 15-year journey.

The Pudneys milk 320 cows on 100ha at on a 485 metres above sea level north of Rotorua. Annual rainfall is 2.7 metres. Keeley, who now manages the herd, moved from contract milking in 2016 to sharemilking with his partner Verena and his parents Mark and Judy in 2022.

Kiwi Fertiliser started working with them from 2012 using Albrecht's method for balancing soil cations. They have a good relationship with consultant Alistair Dagg. The first soil test recorded pH of 7.5, revealing the equivalent of 15 tonne/ha of lime applied when all it needed was one tonne. That overdid calcium and pushed magnesium to half its ideal level, leading to a recommendation of one tonne of magnesium sulphate in 2012 to correct. Also added was boron, sulphur and potassium.

Prior to Kiwi Fertiliser, original recommendations came from leaf analysis instead of soil tests and very cheap Kinleith lime products. Luckily, calcium leaches quickly over several seasons and now the pH is 6.3-6.4. Sulphates (SO_4) usually grab soil calcium first but will also latch onto sodium.

Calcium reduces the availability of potassium. Poor pasture growth was due to low potassium, and now chicken litter addresses that. With potassium prices currently high, chicken litter is economic. Keeley prefers potassium sulphate but potassium chloride is cheaper. With double the annual rainfall in recent years, chlorine easily leaches. Since 2012 no phosphate has been added.

Initially cows got bloat with Kiwi Fertiliser, but not now. They had only three cows with milk fever this season.

As animal health improved so did milk production. Milk solids lifted from 77,000kg in 2017 to 146,000kg 2022 season, from 260kgMS/hd to 450kgMS/hd. This production is on the back of feeding PKE, tapioca and maize. Supplement was usually 70% tapioca and 30% PKE but is now all PKE at 2kg/head.

Keeley is interested in Halter but thinks it is too expensive. He thinks it would be useful for animal health but the somatic cell count is currently 70,000 and has been as low as 38,000. Current MUN is 9-12.

They've been all-grass for eight to nine years because crops yielding 7-8 tonnes provide no advantage. Instead, they over-sow with clovers, ryegrasses, millet, chicory and peas and focus on grazing management.

> By delaying grazing of some multi-species paddocks out to 45-60 days, they can shorten recovery for non-diverse pastures, especially ryegrass stands, to 30 days to keep quality – Russell and Charlotte Heald.

They use longer rounds to reduce fertiliser applications and weed problems. Keeley sees farms that are on 14-day rounds and that apply urea as cows leave each paddock only to end up full of daisies. He is currently on a 29-day round and is keen to push it out to 50 going into mid-summer. The longest round tried so far is 130 days where, strangely, lucerne emerged. This was from March onwards to calving.

Their herd is mixed with Ayrshire, Brown Swiss, Jersey, Milking Shorthorn, Friesian, Fleckvieh and Aussie Reds. Keeley is looking for a cross that puts on and holds condition and is quiet and productive. Fleckvieh is the leading choice.

Of their 80 calves, half are Fleckvieh and the rest Jersey, Friesian, Charolais, Brown Swiss and Ayrshire. They'll milk Charolais just like Galloway. They are milking one Galloway but the cups go on and off quickly. Everything is weaned at 100kg liveweight.

The breeding policy is to target fewer, bigger cows that produce more. Keeley's ideal is ticking down to 250 head squeezing out 500kgMS, though he will need supplements to achieve that. The farm effluent system is consented for up to 350 head but that would require more staff. They're licenced to farm till 2032, so the clock's ticking to push efficiencies.

Keeley's reduced the calving season from 15 weeks to 9.5 weeks last year. He's aiming for eight weeks next year and then six weeks after that with 80% of the herd calving in three weeks. To achieve this he's culling late-calvers, increasing replacements and dropping total numbers. He's also considering short gestation bulls for the last two weeks. Another idea is to calve 10-15 days later as pasture growth often lags due to altitude and lower temperatures.

Keeley likes fully feeding cows and plans to have cows fat by dry-off at end of May as putting condition on over winter is too late. He knows calving earlier is easier with lower numbers, meaning less supplement if pasture growth is slow.

Pumping cows with carbohydrates produces results. He had the highest heifer mating rate feeding them tapioca. They've enjoyed their highest initial calf rate, 75% last season. The empty rate is 12.5% after 9.5 weeks. The bull comes out of the herd on December 20.

Nitrogen budgeting is an issue in the Lake Rotorua catchment. They've planted 30% of pasture in plantain, which acts as a diuretic to flush N out of the body in lower urine concentrations. This means cows require more water. Their neighbours are 100ha of forestry and a Christian camp and they provide a nitrogen buffer.

Changing their fertiliser regime has seen big changes. Not only can they dig into soil now, but more grass and better animal health is allowing them to change their reproduction policy with confidence. Keeley smiles when he says, "Doing my best to farm like the old fellas, but with technology."

Eco-profit Observations

- Property burned out by nitrogen causing severe animal health problems.
- Leaf analysis oversupplied soil calcium levels. Easily fixed as they have a high rainfall.
- Sulphates usually grab soil calcium first, then sodium.
- High calcium prevents uptake of potassium slowing pasture growth. Chicken litter addressing that.
- No added phosphate since 2012.
- Milk production doubled in five years on the back of carbohydrate supplementation.
- Gone all-grass as crops don’t yield enough for ROI.
- Longer rounds reduce fertiliser applications and weeds.
- Shifting to Fleckvieh and wanting fewer, larger cows in herd.
- Looking to reduce calving season from 15 weeks to six, but will require more supplement if spring pasture growth slow.
- Highest in-calf rate with heifers by feeding tapioca.
- Plantain pasture stimulates flushing of dietary nitrogen.

33. Blair and Penelope Drysdale - Vat Volume Priority

In starting out, Blair and Penelope Drysdale's dream certainly wasn't to have cows milking under 300kgMS/hd in their first year. When primary focus is what's in the vat, it is time to get real.

The couple milk 320 cows on 185ha at Te Miro on the edge of the hills west of Norsewood, Hawke's Bay. Their farm came up in a mortgagee sale in 2018 and they managed to beat out sheep and beef producers to get it.

Its history was as a high-input dairy farm built around nitrogen (600kg/ha urea) but investing in top-quality infrastructure made the bank jittery. Added to that, Horizons Regional Council was pressuring high-input operators with compliance rules and the farm had no intensive land use consent.

If a single Waikato herd got bird flu, it would be shut down within minutes of notification. Dealing with extreme business risk necessitates simplicity – Neil and Glenda Gray.

Previously, Blair and Penelope worked for sheep and beef farmers and for consultants owning dairy properties. Their last role was 50/50 sharemilking with 260 cows. Before that they had 400 cows and were lower-order sharemilking, a step down from 800 cows previously. Industry advice to pour on super gave them production problems in their sharemilking jobs. They ended up with Olsen P over 100 and cows doing 290kgMS on twice-a-day milking. Expensive and frustrating.

Penelope has no dairy background and with that open-mindedness ended up at consultancy Outgro. The Drysdales dabbled in regenerative techniques but really started pursuing biological farming five years ago. With feeding cows better to get their nutrients from pasture rather than supplements, animal health improved and inputs reduced. They saw what Hamish Galloway and others were doing with fine-particle applications and pasture mixes. It works.

They started doing it at the 50/50 job at Pahiatua and Blair admits Penelope drove the idea. As pasture quality lifted and the farm improved, confidence grew. Clover increased grass production, promoting chicory and plantain.

Milk production rose by 50-60kgMS per cow and then doubled the previous manager's production, despite being a hilly farm.

Then they bought Te Miro and started facing different problems. Land function was marginal, without earthworms or clover. Moving from mineral-rich Pahiatua clays to free-draining, high P retention volcanic soil dropped milk production by 100kgMS per cow. That was the start.

The first signs of what to come was animal behaviour. They arrived to high-potassium pastures, bloat and milk fever. Cows went crazy during that first year. Feeding stock lick and a lot of salt to lift sodium helped calm cows down. It was frustrating as production was down when the payout was up. Looking around, they realised conventional advice wasn't going to work.

This was when their organic journey started. Influencers were dairying families, the Healds and Meeks, and they realised that they owned their own farm and could do it their own way. Organics fitted with farming and family values and they knew it was a better way both financially and for soils and ecosystems. It didn't take a big shift, other than changing Blair's opinions on weeds and animal health.

But, as Blair points out, milking cows becomes boring doing same things every day. Going organic is made satisfying by teasing out problem-solving skills to try out tools and techniques to make new systems work. Replacing urea with fine lime and seeing no production change meant they were on track.

Furthermore, organic premiums at $9.55/kgMS, about $2-$3 above the conventional price, allowed them to afford fertilisers. Fertiliser is blended off-site with lime, guano, RPR, traces and humates. Then they add Biostart fish, and magnesium sulphate as this can't be blended in crystal form. They wanted more lime to shift soil pH from 5.6 to 6.4 and they have finally achieved that.

To reduce costs, they bought a Tow and Fert to apply smaller loads more frequently. The current application is 350kg/ha, delivering the equivalent of two tonnes of lime annually. They also apply 30-40kg/ha of guano and 30kg/ha of RPR as results are better than each one on its own which is significant with high P retention soils.

Grass grub took out more pasture than drought as pulling synthetic fertilisers during an especially bad infestation left them with cocksfoot-dominant pastures. They've pressed their children into building bird boxes to put them around the farm for starlings with the expectation they'll consume 40% grass grub at no cost.

Cocksfoot-dominant pastures didn't provide enough energy for cows to milk on and multi-species pastures have proven valuable. They've found vat levels differ by 500 litres between the best and worst paddocks over two days. Multi-species pastures provide more flexibility as they can leave swards to recover longer than ryegrass pastures. However, summer rounds don't extend beyond 50 days as production crashes.

Yet, growing good young stock remains a battle as lung worm badly affects them. Through monitoring they found Rescue Remedy didn't work and they achieved better results by drenching with herbal digestives at full moon.

Coccidiosis is also a problem as they can't give coccidiostat and retain their organic status. They are now feeding ad-lib milk by slowly dropping the calf daily intake from 15 to 5 litres by late January when they are weaned. Humates and stock lick also help the calves.

This process cost $50,000, about $20,000 more than milk powders. However, with empty rates 4-7% they are raising fewer replacements and saving money. Blair believes they'll pay their way in their first year milking.

> He had four empty cows (from 280) after a nine-week mating last spring (2023) which coincides with milk urea nitrogen (MUN) in single figures – Henry Hendricks.

Breaking in a new property, especially one that has been abused is a battle. The struggle with many in this situation is how to find root causes of problems, particularly if they contributed to previous owners leaving. With production climbing over 350kgMS/cow, the Drysdales are seeing rewards in the vat for trying a different path.

Eco-profit Observations

- Bought a high input dairy farm built around annual nitrogen applications of 600kg/ha urea.
- Pouring on super at previous sharemilking job created Olsen P over 100 and cows doing 290kgMS on twice-a-day milking.
- Feeding cows better to get their nutrients from pastures rather than supplements, animal health improved and inputs reduced.

- New property had no soil function, high P retention and cows dropped 100kgMS in first season.
- High-potassium pastures, bloat and milk fever resulted in highly upset cows. Salt helped calm behaviour.
- Biggest challenge shifting to organics was Blair's opinions on weeds and animal health.
- Replacing urea with fine lime and seeing no production change meant they were on track.
- Applying 30-40kg/ha guano and 30kg/ha RPR generates better results than each on their own.
- Children build bird boxes for starlings as they'll consume 40% grass grub for free.
- Multi-species pastures provide more flexibility as they can leave swards to recover longer than ryegrass pastures.
- Battle with lung worm. Achieve better results by drenching with herbal digestives at full moon.
- Address coccidiosis by feeding calves ad-lib milk, humates and stock lick.

34. Tim Chamberlain - Pushing Limits

Running late and racing towards a swirling vortex of mist from a helicopter spraying beans, Rose started climbing out of the car window at 50mph. Then it dawned on Tim how passionate she was about avoiding agrichemicals. Only months earlier in London they'd both decided to "have a crack at life", as Tim puts it. Now home on his family's traditional mixed cropping property, the love of his life was demonstrating how to push limits.

Always curious about the natural world, his relationship with it and the art of creating possibilities, that 1984 event started him on an adventure. He was sure there were lots of people like Rose around, so how could he farm in such way and get really good at it? Everybody coming up the drive had a better car and knew more about their business than they did. How do we get to that? he asked. How can we turn it around?

The family farm at Leeston was small, 140ha. Tim's father, Peter, benefited his whole career from agrichemicals and was terrified that Tim, at 25 with no money or managerial experience, wanted to change everything. Tim recalls very loud discussions telling him how to farm his father's land and as locals know, Peter is a considered man.

After disappearing down south for a few days, Peter came back with an idea - put the hardest part of the farm to irrigate into organics for five years, the rest would stay as it was. This reduced initial risk while determining whether organics would work. If not, they could sell up and try somewhere else.

Then a Lincoln University lecturer appeared with enthusiastic students asking Tim, why go into organics. Tim remembers kicking dirt, scratching around for inspiration, when his father said, "I'll tell you why. Five people in this area have died from cancer and we do not know the cause of this." That statement was a bolt from the blue. Tim smiles when he says, "I'm such a fan of my father."

This was during the 1980s rural crisis when interest rates were at 23%. It was scary to be trying something very few knew anything about. So how did he start?

The single best decision initially was trying a little bit of everything on the organic part of the farm - cereals, vegetables and livestock. Without a lot of

> What makes this achievement remarkable is the lack of fertiliser and agrichemicals and little pasture renovation in 40 years – Cliff Shearer.

industry support this reduced risk and built experience. Two pivotal individuals were lecturer Bob Crowder and enthusiastic student Nigel van Dorsser with his major farm management report on converting the farm to organic. If they were on the other side of the Rakaia River, any long conversations would've tallied up as toll calls.

In focusing on growing crops, he hadn't thought about marketing. Peter asked Tim who he was going to sell his Arawa wheat to. Tim had no idea. He'd never sold anything other than to agents. Then organic agribusinessman Jon Manhire turned up and said he could sell it, not bulk but in bags. Peter reluctantly taught Tim to sew bags and they loaded up Jon's Kingswood car to where he could barely see over the dashboard.

Timing is everything. Processing company Wattie's got into organic peas and over five years took every pod Tim produced. Weeds were less of an issue because this crop didn't dry in the paddock. Its success underpinned purchasing the family farm.

Bob's influence increased when he became an inspector for organic certification agency Biogro. His position allowed Tim to visit organic farm inspections with him. He guided them into growing local market onions and carrots. Tim began working with local distributors like Fresh Direct, Ceres, Chantal and others. He started with 1.5m beds, David Brown tractors, then Farmalls, then GPS technologies when working crops, all overseen by English woofer Merf, whose friendship over 30 years consolidated production gains and machinery development.

By 2000 they'd bought and converted the whole farm and opportunities kept coming. They began leasing property at Banks Peninsula for livestock, then leasing ground at Ashburton when local seed company Midlands asked Tim to grow organic carrot and beetroot seed for northern markets. The crop took up 40ha and involved five different farms south of Christchurch.

Then in 2008 he bought Mt Cass Station with two other families during the GFC, a property surrounding Kate Valley, Christchurch city's landfill. Tim got jittery about being stretched so asked Peter to look over the whole operation. It took three days to get around all the enterprises, "just hold in there it'll come right." It was dry and lamb dropped to $3.30/kg but organic lamb had climbed to $6.10 on three-year contracts.

> The beauty of working with nature is that labour can be reduced through genetics. An unspoken challenge is creating an animal that anybody could start organic farming with – Daniel Wheeler.

What made the Mt Cass purchase fortuitous was that the landfill company didn't hand over clear title for 23 months. This delay allowed them to lease before buying at 7% interest instead of 11% two years earlier. Tim's organic reputation opened more landfill company doors. The company was happy to sell and lease more land, and added houses to help with staff. He was swimming in a different current.

Tim's drawn to people, their ideas and enthusiasm energise him. This is what he likes about the regen community - it's attracting adventurers and creating journeys into areas long forgotten or given up by conventional. However, without a genuine brand he's concerned for its potential to become a greenwashing fad.

When asked about dealing with business partners, Tim likes Rick Cameron's advice – "You need to be totally honest all the time, be more interested in the other's outcomes than your own, be totally responsible for everything you do and you have to know when enough is enough."

Eco-profit Observations

- The love of Tim's life demonstrated how to push limits.
- Challenging his father Peter led to changing practices on the hardest part of the farm to irrigate. This reduced risk.
- Tried a little of everything to reduce risk and gain experience.
- Local support crucial to maintain enthusiasm. That only required close confidants.
- Fortunate Watties started processing organic peas. This foundered success of the entire operation.
- Diversified crops and new technologies improved crop performance.
- Started seed production as experience and opportunities allowed.
- Spearheading organic production led to more expansion including livestock.
- Regenerative Agriculture needs a brand to prevent greenwashing.
- Be totally honest in all your dealings with business partners.

35. FINDING BENEFICIAL RELATIONSHIPS

There are a lot of what I'd call industrial farmers who have an ecological conscience, but they don't really make the connection between ecology and complexity and how that relates to their business. And we've found that our business has gone a heck of a lot better since we've been making decisions towards the environment.
David Marsh, 2018 Australian National Landcare Individual Award winner

What's an innovative way to teach children the relationship between ecology and economics? How about sending them into paddocks to gather up all the pesky grasshoppers, put 13 in a plastic bag with a piece of potato to keep them moist and sell them out the gate as Huckleberry Finn fishing kits at $5? This is a classic example of how the late Bill Burrows demonstrated win-win relationships. It took three years to establish a market but it grew.

Allan Savory's Holistic Management challenges people to look for patterns, trends and arrangements to create multiple benefits. Bill, one of the best advocates, said Holistic Management flipped his perspective from cattle producer to sunlight harvester. That opened his mind to monetising second-hand sunbeams, but grounded within his family's values.

Different plants provide different nutrients. If the livestock mineral bill is high, it signifies a lack of pasture diversity – Bland Pastures create Bland Food.

Bill ranched 3,500 acres near Red Bluff, northern California. Despite being large, it could only run 150 head of cattle, whereas he needed 500 to make a living. So he taught at Shasta College for 35 years. There he set up a holistic resource laboratory for students to learn relationships between livestock, cropping and budgeting while challenging traditional thinking.

On 100 acres of cropping land in 42 paddocks, he pulled together 20 pasture-farrowing sows, 35 cattle, 50 sheep and 30 chickens and grazed them as one mob to demonstrate financial, social and environmental benefits. Over nine years, the resource lab's income lifted $12,000, and expenses dropped $10,000, rendering a $14,000 profit above initial 1987 figures. It coincided with increases in soil organic matter from 1.7% to 3.2 %. Using animals rather than mechanical harvesters reduced fuel and labour costs while the sheep/pig/cow/chicken herd provided better utilization of forage than single-species grazing.

He and his students selected breeds and genetics to fit this system, as opposed to maximising production and then adapting management to match the animals. They selected pigs which didn't root up ground or eat baby lambs at birth. They found sows could farrow on pasture and make a bed from a single bale of hay. They didn't need farrowing crates or interventions other than castrating males and using humane nose rings.

While chickens moved with the herd during daytime, they roosted in a nearby eggmobile at night. Hens scavenged waste seeds from various grazed crops and scratched open cattle dung searching for insects, thus consuming cattle parasites. With three different kinds of hens, each laying a different coloured egg, they marketed "rainbow eggs" with cartons containing four white, four blue and four brown.

With larger mobs, paddocks can stay large, reducing subdivision costs. It also makes culling easier as environmental pressures weed out stock not worth breeding from. And it grows more grass – Leveraging Time Lifts Profits.

So how did animals benefit each other? Any animals wandering off were naturally culled as predators were uncontrolled. Sheep concentrated in the middle of the herd as protection from predators that pigs fended off. Cattle learnt pigs broke pumpkins open so they'd stick close to chew on those. Pigs eat cow and sheep dung and benefit from the predigestion while cleaning up ruminant parasites. All these behaviours increased the herd's resilience and reduced costs.

Unable to make a living from ranch production, Bill organised a community barbecue and with the help of friends, neighbours and others, brainstormed a list of enterprises. As a result, 70% of ranch income came from supervised outdoor experiences. He said, "They could have made money from a thousand enterprises but chose those which suited their family's quality of life." For example, with hunters, they banned those wanting deer trophies.

At the forefront of these enterprises were customer relationships. Bill would organise his "guests" to hunt ground squirrels, a pest on his ranch fed to wild pigs. Then, weeks later, hunters would shoot wild pigs. He used one pest to steward the later hunting of another while people paid to control his vermin.

All hunters visiting Bill's ranch were handed a small paper envelope full of seeds and asked to leave it somewhere in the landscape. This was how he spread plant species across his land by taking advantage of roaming hunters.

When he wanted to measure ecological improvements on his land, especially from livestock grazing, he enlisted the community. Instead of doing it himself, he set up a day for friends and others to visit the ranch, then take them to sites he was monitoring for them to record biodiversity measurements. This leveraged his ability to monitor more sites within a single day, while educating the public about ecological relationships, especially the positive impacts of grazing management. At the end of the day there would be a barbecue and party.

Valuing genuine relationships extended to his commercial dealings. When developing his property with more pasture subdivision, instead of ordering a truck load of troughs, he went to the local cement factory with his own trough mould. He asked returning trucks to pour any left-over cement into the mould and call him when they were full and set. That generated two benefits. Firstly, it created value for cement destined for waste and secondly, it fitted Bill's cash flow better by reducing capital tied up in farmyard inventory.

Bill was utterly committed to his family's values. Sometimes he was so busy, he and wife Kay would communicate through a feelings diagram on the bathroom mirror. As their children reached adulthood and scattered across North America, they renamed the ranch. In calling it The Resort, they built a chalet for their children to bring their families for extended visits and keep them involved in ranch decisions.

Holistic Management flipped how Bill Burrows saw himself. By drawing on family values for commitment and courage, his actions demonstrated ingenuity and integrity by creating multiple benefits within his community.

Eco-profit Observations

- Selling grasshoppers as fishing kits taught children links between ecology and commerce.
- Flipping from cattle producer to sunlight harvester opened his mind to monetising second-hand sunbeams, but grounded within his family's values.
- At Shasta College he set up a laboratory for students to learn relationships between livestock, cropping and budgeting while challenging traditional thinking.

- By combining pigs, cattle, sheep and chicken to graze as a single mob on 100 acres, profit lifted $14,000 in nine years and nearly doubled soil organic matter.
- Selected breeds and genetics to fit this system, as opposed to maximising production.
- Marketed "rainbow eggs" with cartons containing four white, four blue and four brown.
- Sheep concentrated in the middle of the herd to avoid predators that pigs fended off.
- Quoted saying, "They could have made money from a thousand enterprises but choose those which suited their family's quality of life."
- He used one pest to steward the later hunting of another by people paying to control his vermin.
- Paying guests spread plant species across the property while hunting on the ranch.
- Created a community day to measure biodiversity changes on the property.
- Used a trough mould to reduce cement factory waste and capital tied up in farm inventory.

36. Vince Payne – Productive Pasture Predators

As the saying goes, if you can't beat it, eat it. Dannevirke dairy farmer Vince Payne is easily one of the most enthusiastic and knowledgeable farmers observing predatory insects. He runs a high-profit operation milking 150 cows, once-a-day with farm working expenses around $4/kgMS. He could easily hire staff but, as he puts it, he doesn't want to be chasing his tail all day. That said, he has to feed his cows well so they don't eat ragwort.

Vince bought his 80ha property, Windwoods, in 1995, claiming the worst farm in the best district. His roadsides, pastures and riparian areas used to be yellow with ragwort (*Jacobaea vulgaris*), not now. Within a week of arriving, a Horizons Regional Council weed officer appeared on his doorstep asking how he was going to deal with his ragwort problem. As he sprayed herbicide, he thought there must be better and permanent ways to deal with this weed.

He recalls harvesting maize infested with army worm in Waikato in the early 1990s and yet yields suffered little. Tiny wasps parasitised the worm before it burrowed into corn cobs and in a matter of days saved crops from destruction. This discovery was his spark.

Diamondback moth decimated plants in sandy soil due to moisture stress while plants growing in silt were perfect – Will Perriam-Lampp.

According to Landcare Research, it's estimated over 25,000 species have been introduced to New Zealand in the past 200 years. The number of released weed bio-control agents increased in the last 30 years due to disillusionment with herbicides. The current success rate for establishing weed bio-controls is very high by global standards - some 85%, nearly twice the 1980s rate of 44%. This is due to a biosecurity network that allows the release of large numbers of agents at suitable sites and monitors their progress. Landcare Research estimates the benefit:cost ratio for agricultural weeds is 155:1, for every dollar spent, there's a benefit of $155. With ragwort, it's estimated at 860:1.

Vince knows weeds germinate because of mismanaged bare soil. Research from the 1930s confirms higher post-grazing residuals reduces weed infestation, backed up by more recent research into deferred grazing. As Vince observes, because New Zealand's climate is constant, weeds germinate throughout the season, whereas in continental climates it's snow thaw that triggers germination.

Because a consistent climate is not that common around the world, biological controls find copying this behaviour difficult. That said, certain weeds germinate before others: Wing (*Carduus tenuiflorus*) and nodder (*Carduus nutus*) thistles are earliest, scotch (*Cirsium vulgare*) thistle next, calis (*Cirsium arvense*) are later again. As Vince's climate dries and warms, scotch thistle numbers increase.

For 30 years he's passionately advocated for farmers to develop skills identifying phases of plant growth because insects predate plants at different stages and ages. It means not only knowing when to look for predators but where to look, whether it be in the roots, stems, leaves or where leaves join stems, and to look for tell-tale signs like frass (insect dung), eggs, puncture marks and grazing patterns on leaves. He uses Google Lenz to identify insects he finds on plants.

Vince, a pilot, flies to places to pick up pasture predators, using a leaf blower to suck up caterpillars and insects into a stocking. Now he has an army of Windwoods Warriors, bio-control agents for dealing with problem plants. Some familiar thistle predators on his place include Green Thistle Beetle (*Cassida rubiginosa*), Scotch Thistle Gall Fly (*Urophora stylata*), Vagrant Twitcher or Small Thistle Moth (*Tebenna micalis*) and Thistle Receptacle Weevil (*Rhinocyllus conicus*). These agents attack scotch and Californian thistles at various stages, either by eating the foliage or seeds. He believes 70% of thistles succumb to insect predation on his place.

With a reputation as the local ragwort farm, he has three predators to deal with it, Cinnabar Moth caterpillar (*Tyria jacobaeae)*, Ragwort Flea Beetle (*Longitarsus jacobaeae)* and Plume Moth (*Platyptilia isodactyla)*.

Pastures riddled with grass grub provided little winter grass, making for a tough first spring – Peter and Christine Johnstone.

Horizons bought plume moths from the West Coast and Vince's property did the first local trial. Six years later 80% of ragwort is predated by plume moth due to its longer emergence period. It appears by January when Cinnabar caterpillars are finishing damaging flowers and is well established as flea beetles arrive 20th February.

His farm experienced its largest ragwort germination three years ago from porina (*Wiseana cervinata*) and grass grub (*Costelytra zealandica*) damage. Incidentally with porina, a forest fungi (*Ophicordyceps robertsii*) increases with populations and grows up through porina moth and turns it into a vegetable caterpillar. Historically, Māori burnt these caterpillars and used its ash as ink for tattooing facial moko.

In fact, Vince's property has become a haven for bio-controls with many different insects found there, Irish wasps (*Microctonus aethiopoides*) which predate clover root weevil, (*Sitona lepidus*), harlequin ladybirds (*Harmonia axyridis*) to deal with giant willow aphid (*Tuberolachnus salignus*), broom seed beetle (*Bruchidius villosus*) to deal with yellow lupins in his waterways and a gall mite (*Aceria genistae*) to chew through broom, to name a few. Not all problem plants have biological controls. Cattle trampling deals to giant buttercup by crushing its crown.

As a champion of bio-controls, Vince is wary of using herbicide. While spraying next year's weed crop at rosette stages only requires quarter the chemical, this is also when insects attack these plants. Field trials demonstrate herbicide kills more insects than insecticides because it's non-selective. Lower predator numbers increases weed-spraying in later seasons.

Observing ecological predators is a useful skill to help farmers take an interest in their surroundings and learn about environmental health and function. These insects graze plants ruminants avoid and reduce use of synthetic poisons. The beauty is they are self-organising, self-repairing and, best of all, self-replicating, thereby minimising cost and labour of keeping troublesome plants under control.

Eco-profit Observations

- Buying the worst farm in the best district started his journey into biological controls by questioning the value of spraying herbicides.
- Success rate for biological controls in New Zealand is very high due to the national biosecurity network.
- The cost:benefit ratio for farm weeds is 155:1, and for ragwort 860:1, according to Landcare Research.
- While snow thaw in continental climates trigger weed germination, New Zealand's temperate climate makes weed control more challenging for biological predators.
- Uses a leaf blower to suck plant predators into a stocking.
- Believes 70% of thistles are predated on his place.
- Bare soil from grass grub and porina stimulates ragwort outbreaks.
- Using herbicides decimates predator populations because the chemicals are non-selective.
- Identifying weed predators is useful for pastoral ecosystem health.

37. Dennis Nieuwkoop - Optimising Land Income

Dennis Nieuwkoop always wanted to farm. To raise capital he started shearing and bought a three-acre lifestyle block. Over decades, using a range of enterprises and initiatives, his holdings increased to 200 acres, along with machinery, animals and people, to reach his goal. His career is a lesson in creating complimentary enterprises.

Starting with small acreages, he used simple income streams to create flexibility. Growing cattle gave him opportunities to search for other income sources.

Working for various seed and produce companies introduced him to potential future clients. That got him into spray contracting, resulting in buying a property with soil so toxic from growing vegetables, it didn't grow a blade of ryegrass for 12 years. This was the start of looking for alternatives.

Taratahi Agricultural Training Centre offered Dennis opportunities to teach and he developed the rural contractor course. That led to contractors testing equipment at his place, so he used those occasions to subsidise machine labour. The experience introduced him to possibilities where both parties can benefit without a sense of being used, a business insight reinforced when meeting Joel Salatin years later.

These experiences, plus those of clients, got him into spraying biologicals. He's found putting on nitrogen, fulvic acid and fish at 20 l/ha improved old, dense pastures once fed a little. Why reach for the plough when that works? It pushed him into no-till where later he found a secret to good germination on heavy clay was aeration.

But in a surprising twist, he fallowed several paddocks and realised these paddocks were the only ones he didn't resow – Peter Coles.

That created another enterprise - his own crop programme for farmers. For clients wanting a winter crop with brassicas and turnips, he made Jungle Mix, mostly with tap-rooted plants like radishes, clovers, chicory, plantain, sorghum to use root systems to break soil pans. Other mixes included Soil Builder, Summer Feed and Reboot using 29 species to morph into a permanent dairy pasture costing $600/ha which narrowed down to 6-8 species after a couple of years.

Mistakes from complicated recipes foster simpler practices. Upon spraying winter crops with Round-up, residual meta-sulphide in the tank smoked all the herbs and legumes in a winter mix, only grasses emerged. That got him renovating pastures without herbicide by severely grazing first and then direct-drilling chicory, plantain, radishes and white clover. A week later he'd regraze with cattle for a day to remove flag leaves. He didn't sow ryegrass as it came up in the second year anyway. By working more with nature, he was getting better results cheaper with low risk.

Such mixes are used for balage. While he's found making and feeding balage is expensive and less risk in spring, making hay produces more consistent pasture growth.

When he met Joel Salatin and his philosophies on natural approaches and cycles, his claim of making $78,000/acre income was inspiring. How could that happen on 12 acres? A comment from Joel got him thinking farming people might be more lucrative than animals.

He turned his American barn into two Airbnb suites. Then developed a caravan park and eventually built tiny home sites. Farming people generated $80,000-$90,000 a year. Of the 11 families paying to be there, many were immigrants or solo parents trying to get a start. His most successful year generated $175,000, including other income from his seed business, ag teaching and first-aid training. Outside income paid off his property in 10 years.

> Cameron Forbes suggests an enterprise for entrepreneurs; breeding sniffer dogs to detect fly-prone animals.

So, what other possibilities come with farming people? One is land sharing where owners bring their tiny homes without needing to buy land. Their rental is $150 a week for power and water.

In looking for mutual benefits with his tenants, Dennis began growing vegetables, fruit trees and pastured chickens to teach families how to grow food. In creating a supportive community, tiny home owners swapped their labour to barter for vegetables or meat. By developing ownership in food-growing, this also established security. Dennis reasons land-sharing using tiny homes creates a stepping stone for owners to eventually secure their own places.

His latest project involved buying a small dairy runoff which had a decent shed but no electricity or water. Once electricity was connected, he rented the building to an automotive business.

Another possibility is a viewing platform for people to watch aircraft flying in and out of the local airport. Their low approach over the property is why he lives somewhere else. Air Force helicopters once flew so low they blew a transformer off a nearby power pole, hence the possibility for entertainment.

Furthermore, a surprising lesson in reducing infrastructure costs came from sinking a bore well.

The diviner said water was at 24 metres. A friend turned up with three-metre lengths of 20mm steel pipe. The first pipe into the ground had a chisel at one end while the top was connected to a high-pressure hose to push soil up outside the pipe lining. They simply twisted it deeper, adding three metres at a time till they hit 24 metres where the pipe almost disappeared hitting a cavity. The gap created around the pipe allowed insertion of a 30mm pipe with a foot valve. The pressure pushed water to within 4 metres of the surface allowing a small Davis piston pump to draw it from there. Total cost was 4% of the original $72,000 quote.

Dennis' radar for business opportunities is highly tuned. He proves working with people can be every bit as profitable as working with crops and livestock. Creating complimentary enterprises is another way to farm ownership.

Eco-profit observations

- Bought a property with soil so toxic from growing vegetables, it didn't grow a blade of ryegrass for 12 years.
- Contractors testing equipment at his place subsidised his machine labour, a benefit where both parties didn't feel used.
- Nitrogen, fulvic acid and fish at 20 l/ha improved old, dense pastures, removing the need to plough.
- Aeration is the secret to good seed-germination on heavy clay.
- Developed his own seed mixes to create ecosystem services, for example sowing tap-rooted plants to break soil pans.
- Renovated pastures without herbicide by severely grazing first, then direct-drilling pasture mix, then regrazing with cattle to remove flag leaves.
- Making hay produced more consistent pasture growth than cutting silage.

- Creating a land rental business generated more income per hectare than livestock.
- Getting tiny home owners interested in growing food swapped labour for produce, improved security and developed skills for self-sufficiency.
- Selling a view or experience is a possible income-earning enterprise.
- Simple technique to drill for water.

38. Ross and Karen Potter - Going Bananas

New Zealanders consume 73 million kilograms of bananas annually, all gassed with ethylene. Northland can grow bananas, no chemicals required if managed properly, but a viable packhouse requires 10,000ha of plantations. Till then, there's plenty of scope for those willing to take a risk.

Ross and Karen Potter farmed 300 dairy cows on 165ha near Maungatapere, Northland. After a lifetime of cattle and forestry they were looking for something different. The owner of their leased land agreed to buy their herd and lease their land (60ha) to milk with a manager. That's provided an opportunity to explore banana growing. It's a bit different to managing cows as they charge into the world of direct marketing.

Banana production is dependent on temperature with more production in summer than winter. Northland can be a little cool for traditional varieties such as Cavendish, but Ladyfinger varieties handle cooler conditions better. Shelter is essential in Northland and some farms grow bunches weighing 25kg.

Now they're considering permaculture with eco-design consultants Greenbridge developing landscape plans for shelter belts to reduce wind flow – Kenneth and Rachel Short.

It takes three years to get good fruit. The best production is from three-stem clumps (mother, daughter and granddaughter) as the fruit is bigger. The best advice is to plant 1,000 clumps/ha at 2-3 metres spacing with a four-metre gap for tractors to pick up fruit.

Now (2025) they have 3,300 clumps of bananas with 3-10 stems over three hectares. The reason for so many 10-stem clumps was because of Covid. They had orders for 6,000-7,000 pups but ended up fruiting them instead. Companion plants can include pineapple or watermelon, but diverse cropping systems like perma-dynamics produce smaller fruit due to competition for nutrients.

Bananas have a fibrous root system. If planting in clay soil, ripping the ground first prevents clumps "lifting" in the hard soil.

Creating uniform bunches requires bagging to prevent damage and blemishes from leaves, rain and birds. Avocado frost cloth is ideal as it's breathable and transparent. Ross designed and developed his own bag applicator to slide bags on from ground level to eliminate ladders.

Bags encase bunches for up to nine months but their rubber bands deteriorate in sunlight. Bananas attract all sorts of birds and bird pecks start fruit maturing by stimulating natural ethylene. To stop ethylene production, simply remove pecked bananas.

All bananas are picked green but their tips must fill out properly. From there, ripening speed is managed with controlled atmosphere in a ripening room. In summer, 13-14C will slow maturity by two weeks. In winter, heaters speed up maturity. When exposed to cold, hard lumps develop in the fruit. This can also occur due to mineral deficiencies.

Stems have a centre of pure starch so everything wants to eat it, especially birds like pukekos and turkeys. Hence, Ross uses non-glyphosate herbicides for weed control instead of livestock. Bananas have potential for standing livestock feed in droughts but bird control is essential during establishment.

Growth is phenomenal. In November and December leaves emerge, grow and fill out in three days. A full-sized stem weighs over 100kg. Steep hillside bananas could generate more income than grass, even from pups which emerge from underground corms.

Like livestock, bananas move, mostly north or north-east. Walking of bananas is controlled by removing pups. However, taking too many causes other stems to tip over and collapse. Dairy effluent generates pups as it's full of nitrogen. More pups, less fruit.

When it comes to marketing, hobbyists can easily sell to consumers or to farmers markets but supermarkets require GAP (Good Agricultural Practice) certification. Commercial operators need at least NP1 (National Programme One) to sell edible product. The Potters do both frozen and freeze-dried bananas and have NP2 certification for processing. This is a level below GAP.

> Many will buy other well-known brands thinking they're small operations only for John to tell them they're owned by overseas corporations – John

Postage problems led them to buying a van for delivery runs to Auckland. Karen makes trips once a week, meaning they can ensure banana freshness and quality to customers. They have on-sellers at Wellsford, Matakana, Silverdale, Orewa, Parnell and Grey Lynn markets. Because of local small growers (farming 200 stems), they don't sell at the Whangarei growers' markets but do sell at the Onerahi Saturday market.

Bananas generate other products. Bees love banana flowers so they're ideal for banana honey, which tastes like sweet clover honey and is dark in colour. The Potters believe they can produce around 350kg of honey from 30 hives. Banana leaves also have huge potential, especially if steam-pressed into cutlery. Fresh banana leaves are also valued by local sub-continent ethnicities.

There's also frozen and freeze-dried banana chips. Bananas are frozen first, then freeze-dried over 24 hours, reducing from 3.3kg frozen down to 1.0kg freeze-dried. Frozen bananas are sold to cafes in 1kg packs for smoothies and freeze-dried are sold in 20gram packs as natural sweets.

They're considering expanding to 40-50ha but unsure if they want the hassles of going bigger, such as employing staff. Achieving GAP would push more volume but labour requires harvesting a minimum of 5kg/hr, so they would need 10kg bunches or heavier. Selling 250kg/week from October-March (6 months) requires 6,500kg or nearly 2,000kg/ha over summer. Last summer they achieved 150kg/week, just over 1,100kg/ha. Full-time staff would need 200kg of bananas a week for one person full-time.

Other crops they grow include watermelon, rock melon, honey dew melon, passionfruit and pineapples.

The change from livestock has refreshed and inspired the Potters about farming, now they're riding a crest of a wave. Juggling growth and efficiencies is stress every growing business experiences. For bananas to truly establish in Northland it's going to require meaningful investment and support.

Eco-profit Observations

- Banana production is temperature-dependent. Best to pick the right banana.
- Best production comes from three-stem clumps.
- Companion planting produces smaller fruit due to competition for nutrients.
- Ripping clay soils before planting reduces stem-lifting.
- Presentation is everything and protecting bananas with avocado frost cloth is best.
- Can manage ripening if picked at the right stage.
- As stems are pure starch they can't use livestock for weed control.

- Bananas could make hillsides generate more income than grazing.
- Dairy effluent grows more pups, less fruit.
- Bought their own delivery van to eliminate postal delivery problems.
- Bananas have potential for a range of products from the flower, fruit and leaves. Processing fruit provides more opportunities.
- They're now growing other sub-tropical fruits.
- Juggling economies of scale and labour with demand is challenging.

39. Paul and Prue Ensor - Woolshed to Wardrobe

Paul Ensor returned home from his travels in 2004 to take over the family property, Glennan Station, 1,035ha on the south bank of the Rakaia Gorge with river flats, easy and higher hill country. He was now in charge of 5,000su, including some 2,500 ewes. Stores made $25 for a 25kg liveweight lamb. The business needed improving on many fronts but Paul was willing to challenge the root causes.

In 2006, tenure review brought new growth as they added freehold river flats and a hill to freehold fan country. Higher country went into the DoC estate, bounded at the original snowline. It allowed them to invest in subdivision on the hill face with all 27 paddocks now 15ha and to build a gravity-fed stock water system. River country saw the biggest changes with three paddocks split into 50, allowing better grazing management.

His focus was finding time efficiencies to juggle stock work with development fencing. He gravelled all laneways and built cattle stops instead of swinging gates – Richard Thorp.

Two decades later, Glennan runs 7,500su, mating 1,200 ewes to a Romney terminal sire and 2,000 ewes to Merino sires. The Merino lambs grow faster but they are all kept through winter, shorn as wethers and sent to the works. The move to more meat production has lifted wool micron which has been as low as 13.8 but is now around 16.5. Merinos lambed at 132% last season (2024).

There's also 100 Angus cows and they overwinter R2s by buying in heifers and killing them in spring when premiums are higher. Calves are weaned straight onto multi-species pastures and grow well on sunflowers, an idea that emerged from exploring new pasture systems.

Glennan is split between hill (56%), fan (15%) and river country (29%). An old riverbed was traditionally sown in straight lucerne but funnelling winds stripped topsoil, creating dust problems and fertility loss. With lucerne seeds striking in grass paddocks next door, and seeing ewes in lucerne paddocks eating everything along fence lines except lucerne, it was clear to Paul traditional practices were costing production. Talking with other farmers led him to Sam Lucas at Agricomm and around 2012 they began looking at lucerne mixes.

Lucerne paddocks never get a proper chance to seed, so a two-phase cropping programme reduces weeds and invigorates pastures. It starts with drilling monoculture ryecorn in January to 30mm, sometimes with clover and phacelia; then in spring sowing an annual/perennial mix including sunflowers, various clovers – crimson, red, white – phacelia, bromes, prairie grass, cocksfoot and phalaris at around 25kg of seed/ha. A primary concern is pasture tolerant of grass grub, so no ryegrass. Fertiliser use includes RPR, guano and a lot of dolomite. A crop of swedes is also grown, but on higher fertility fan country to get 15tonnes/ha for winter feeding ewes.

Furthermore, multi-species pastures reduce horehound, meaning less spraying and chemical subscriptions. Weed spraying encourages horehound.

Wind can be troublesome so ewes lamb on hill or fan country as they're too exposed on river flats. Shelter is an issue at lambing. One winter a decade ago they experienced three days of rain followed by 30cm of snow and ended up weaning barely more than 50%. Snow risk is now addressed by staggering lambing from August 20, but most ewes start from September 12 with three mobs lambing a week apart.

Rakaia Gorge river flats are a designated significant natural landscape, which means traditional shelter belts of pines or macrocarpas are prohibited. Without fence line shelter, Paul is exploring other ways of disrupting wind gusts, such as a semi-woody savannah pattern with natives and taller grasses. Cabbage trees have emerged on their QEII Trust block so he is wondering what other natives will appear over time.

A hydroponic operation allowed him to tinker with growing lettuces, herbs and strawberries in Southland's cooler climate while dairying – Robin and Lois Greer.

The importance of working with the land is something Paul and his wife Prue have schooled into their three children. Paul says he has four roles: Dad, farmer, NZ Merino director and entrepreneur. Sustainability values are at the core of their woollen garment business, Hemprino.

Research showing merino blended with other fibres would improve its properties inspired an opportunity for Paul and Prue with two other shareholders to launch Hemprino in 2021. They liked the idea of controlling the whole supply chain from woolshed to wardrobe. They're cashflowing it themselves and learning how to develop a brand during tough economic times.

All processing is done in New Zealand. Blending of hemp and Merino wool happens at yarn stage in Wellington and the garments are knitted in Tauranga. Wool value goes from $20/kg clean to $60/kg as yarn to $120/kg as an item of clothing. The blend is 80% wool, 20% hemp. While wool comes from their own operation, hemp is another story. Despite all the noise in New Zealand for hemp opportunities, they import instead as only China has enough expertise to provide textile-quality hemp. Turkey is next.

One of hemp's properties is it softens over time as wearing flexes its fibres. It also doesn't take colour, creating an attractive fleck. Hemprino is great for wool-sensitive people as it doesn't irritate skin.

Producing a fabric is a possibility but not yet. Jerseys sell for $329, making a price difference with the Icebreaker brand. They avoid discounting to prevent a race to the bottom. In 2024 they moved to supplying retailers after establishing online sales. Everything is distributed from nearby Methven, base of another shareholder.

Defining a niche by linking one's own values to products and communicating benefits to consumers is a pioneering journey for farmers. Moving from commodity producer to direct marketer requires beliefs that production systems matter and that customers do reward entrepreneurs with premiums. Compare that to the commodity world where being different usually requires discounting. That's not the future for Glennan Station.

Eco-profit Observations

- Tenure review provided opportunity to subdivide paddocks, get better grazing control.
- Moving to meat production has lifted micron diameter of wool.
- Livestock slaughtered in spring for higher premiums plus a wool fleece.
- Calves grow well on multi-species with sunflowers.
- Observing lucerne seeds striking over the fence and ewes entering lucerne paddocks eating everything along fence lines except lucerne triggered looking at lucerne pasture mixes.
- Doesn't sow ryegrass as not grass grub tolerant.
- Multi-species pastures reduce horehound meaning less spraying.
- Staggers lambing date to reduce weather storm risks.
- Exploring semi-woody savannah pattern with natives and taller grasses to reduce wind gusts while meeting local vegetation policy restrictions.

- Sustainability values drive garment brand from production to processing.
- Cashflowing new enterprise to keep control while developing their brand.
- No fabric-quality hemp grown in New Zealand.
- Blending hemp and wool great for wool-sensitive people to wear.
- Mindful of discounting as it doesn’t fit with brand values.

40. Simon Lee - Staff come First

Instead of replacing depreciating assets, how about investing in skills and management training? American farming editor Joel Salatin predicts that in an age where technology is becoming the solution to farming efficiency, its replacement will eventually drive farm costs higher than wages and salaries. The highest ROI comes from skills and knowledge. How do staff buy into that?

Glenaray Station, near Gore, Southland, is one of New Zealand's largest properties, covering some 70,000ha of private and crown land while overwintering 94,000 stock units comprising of 40,000 Romneys and Perendales, 2,000 Merinos, 2,500 cows and 3,800 hinds. There are 26 staff, 21 of which operate the farm.

Simon Lee arrived in 2022 from 18 years at Mendip Hills Station, North Canterbury. While looking forward to the challenges of Glenaray's different enterprises, the farm board approached him about a new direction as his appointment created an opportunity to build a proactive culture to support staff.

So, his first action wasn't increasing production, it was aligning the board's vision to a road map for staff development and growth. The intent of Glenaray Station includes staff practicing respect for the environment, people and livestock, as well as creating success for business, staff and community. The board wanted a pathway to align new staff with these objectives while creating a staff culture of continual improvement for the Glenaray "brand" to push its capacity.

So, how do you develop a company culture for the crew? That's where Lean Management became useful. Simon read Jana Hocken's The Lean Dairy Farm, focusing on efficient dairy farming. Changing culture requires all onboard, turning a spotlight on how staff see management and addressing their concerns. Jana workshopped with them for a day in Gore. Shepherds and block managers turned up with three things Glenaray could improve. Among 50 ideas was a dog training area, more time to learn horse riding skills and buying a digger.

Jana's workshop and visits initiated a culture of continuing improvement by blending company culture with health and safety and infrastructure. Everything has its place to lift efficiency.

One aspect easily incorporated into routines and behaviours was health and safety. Examples included an armoury for storing shepherds' weapons and parking all utes next to the shepherds' quarters.

Transparency and accountability were strengthened further with all staff on Whatsapp channels, using Resolution for sharing documents and Monday morning meetings for everybody to meet face-to-face. All shepherds rotate around blocks over four weeks to prevent ownership and territorial problems.

Also, investing in infrastructure improves labour efficiency. Glenaray is 80km long with headquarters, workshop and shearing shed at its southern end. It now has four sets of satellite yards, each with its own compressors, auto-drafters, air hoses, etc, so shepherds aren't chasing tractor drivers for gear. That reduces mustering and travel time, resulting in less liveweight shrinkage droving stock to and from facilities.

All other equipment has its place in the main workshop making it easier to find and ensuring nothing is missing when needed. When staff tidy up after themselves, it strengthens pride in working at Glenaray, builds team culture and attracts new staff.

Efficiency is extended to machinery. If it's worth $500 and sitting under a tree, it's out the gate and the cash is put into more productive investments.

> In the early 2000s a pivotal moment arrived when he couldn't pass a medical for his dive ticket. The doctor told him he had to change his job, a wake-up call – Adrian and Pauline Ball

Scale brings opportunities for staff as Simon sees his role preparing them beyond farming life. All get exposure to professionals such as accountants and mortgage and bank managers to build practical life skills and knowledge rather than just trade skills through industry courses. It's who you know, not what you know that makes a difference as life progresses. How many farmers ever map their social network to know where to invest?

However, one criticism of Lean Management is an expectation large operations can run staff on recipes like McDonald's franchises. Simon's view is that's more difficult with drystock than dairy. Firstly, dairy is only one class of livestock whereas Glenaray has multiple classes. Secondly, scale increases complexity, therefore shepherds are best educated on principles instead of strict formulas or routines. Farm staff need freedom to make choices from multiple options instead of being railroaded to one path.

What's the reward for staff? More autonomy to develop self-confidence and resilience. For Simon, he sees staff engaging in a culture of wanting to understand the why. That's lengthening time staff stay at Glenaray.

Grass grub is a big problem for all local farms and Travis' previous experience with fescue presented a logical solution – Travis Leslie

When staff are engaged, it frees up time for Simon to explore opportunities with Glenaray's resources. For example, harvesting 2,800 wild deer for venison, helicoptering tourists to overnight huts for snowmobile experiences and a deer, possum and ram-based pet food. Only through a reliable culture can Simon recharge his batteries and look after his own mental health.

While returns from staff development grow, Simon believes Glenaray can overwinter 110,000su without further pastoral development, just utilising what they have better. For example, by staggering lambing dates they wean all lambs at 82 days resulting in 80% done before Christmas. All lambs are finished on 780ha of summer brassicas, mostly raphno, and new pastures. That's jumping production from 5,000 primes to over 31,000 in the 2024/25 season, around 60% of lambs weaned.

As ewe numbers climb and lambing percentage lifts from 132% to 140%, it'll produce another 9,000 lambs to reach 40,000 primes through better feeding management.

Burke Tiechert, retired US ranch economist says, "Leadership isn't taking charge; it's taking care of those in your charge. It's measured by voluntary responses of others when providing good motives to follow." It requires a degree of vulnerability and that's how Simon's humility is accelerating production and financial returns from Glenaray's fixed labour costs.

Eco-profit Observations

- Highest ROI always comes from skills and knowledge. How do staff buy into that?
- Looked for pathway to align new staff with board objectives while creating a staff culture of continual improvement.

- Changing culture requires all onboard, turning a spotlight on how staff see management and addressing their concerns.
- Transparency and accountability are strengthened with all staff on Whatsapp channels, using Resolution for sharing documents and Monday morning meetings for everybody to meet face-to-face.
- Glenaray has four sets of yards, each with its own compressors, auto-drafters, air hoses, etc, so shepherds aren't chasing tractor drivers for gear.
- When staff tidy up after themselves, it strengthens pride in working at Glenaray, builds team culture and attracts new staff.
- If gear worth $500 and sitting under a tree, it's out the gate and the cash put into more productive investments.
- All get exposure to professionals such as accountants and mortgage and bank managers to build practical life skills and knowledge rather than just trade skills through industry courses.
- Scale increases complexity, therefore shepherds are best educated on principles instead of strict formulas or routines.
- What's the reward for staff? More autonomy to develop self-confidence and resilience.
- Only through a reliable culture can Simon recharge his batteries and look after his own mental health.
- Glenaray can overwinter 110,000su without further pastoral development, just utilising what they have better.
- Leadership isn't taking charge; it's taking care of those in your charge. It's measured by voluntary responses of others when providing good motives to follow.

41. EQUITY CRISIS LESSONS

There is only one bank in Monopoly and it doesn't have any pieces on the board.
Johnny Moscato

Before October 20, 1987, Black Tuesday's sharemarket crash, over 90% of non-performing and foreclosure bank portfolios were farm debt. Within a week it was less than 1%. Overnight equity companies evaporated with worthless assets and the responsibility to edge New Zealand forward fell on farmers, much like the covid crisis.

University of Otago Professor Hugh Campbell interviewed 30 Ashburton farming families and 30 rural businesses for his PhD after the rural crisis in the early 1990s. Today's generation has long forgotten events of the 80s due to a decade of cheap money. No other OECD sharemarket reached such heights, nor plunged so low and took so long to recover than New Zealand. Those who never lived through it remain blind to its lessons - return *of* capital is more important than return *on* capital.

> Nation said this was why small business operators should never take advice from Wall Street managers because they removed themselves from the responsibilities of debt – Consultants Breed Costs.

The dawning of Rogernomics (Roger Douglas, Minister of Finance 1984-1988) saw bank services heavily restricted, especially in what they offered farmers. Securing loans was hard but banks were not exposed to risk. When the economic model started collapsing, politicians smashed the financial protocols, launching New Zealand's economy onto a roller coaster that still impacts today. New regulations opened up the finance sector in 1985, igniting the sharemarket's meteoric rise, and banks, free to borrow from overseas, became drunk on newly discovered freedoms. For example, Bank of New Zealand bankrupted itself by lending to clients investing in over-inflated sharemarket gains. The financial sector lost sight of New Zealand's economic foundation.

With political leaders stating farming was a sunset industry, coupled with an elevated dollar reducing commodity prices, farming tanked. Banks giddy on rising share values targeted the rural sector to improve returns. Land values collapsed as nobody, including banks, wanted farms because of low returns compared to the sugar rush of sharemarkets. As farmers struggled, banks cleared out long-established account managers, people embedded in local communities who understood farming and knew the families. Instead, they were replaced with fresh-faced university graduates using a six-step metric to rank desirability of clients. Of the 1,600 farmers in Ashburton County, 127 were issued notices of foreclosure. Black Tuesday forced a halt to the foreclosures and, put simply, that saved New Zealand farming.

If the crash had happened 18 months later, New Zealand agriculture would've been a basket case for decades. Campbell says the sharemarket crash separated the farm equity crisis into two phases, hostile and progressive.

The hostile phase tactics created a hatred of banks stronger than the current resentment of politicians and their carbon crusade. Switching from assessing the farming family to only using land asset values as sole determinants of ROI drew a line between farmers and their support network. It created lingering attitudes among farmers toward many professional communities which have yet to recover. When Australian banks like ANZ arrived in the 1990s they secured waves of new clients.

Before the crash, farmers set about bringing down costs and delaying any potential forced sale. Campbell identifies seven strategies rural communities used to beat back banker bullies.

As banks cracked down on seasonal finance, farmers dropped fertiliser applications and started milking their land's natural capital. Many realised they'd been putting too much on. They pulled back other expenses, delayed development, laid off employees and bought kids home from boarding school. Survival required entrenchment.

Women gave up their social roles and took over bookkeeping or used previous skills to bring in a second income. In the mid-80s it was unusual for women to work, but by 1992 to have both partners working was normal.

Other actions included entering the black economy and selling over the fence. Farmers made sure they were feeding each other.

They also created ways of lifting farm equity values. Because soil type was tied to land value, one father and son worked 48-hour shifts clearing river stones from three paddocks to change soil type and lift their property's value beyond receivership.

Families worked with accountants to hide savings, insurance, legacies, trusts and liquid assets such as machinery from bankers. People forget that this was a time when agents would inspect houses for luxury foods, for example, white versus brown sugar. Any bankers' cars on country roads triggered calls ahead to give others time to lock their houses and head for the back boundary.

To deal with high interest rates farmers began leasing from their own families or moving their lending away from banks back to solicitors' trust funds or equity partners.

Federated Farmers urged farmers to leave stock and station firms en-masse to support local co-ops operating on a cost-plus basis, such as the Ashburton Trading Society. When farms were issued a foreclosure notice, farmers picketed auctions so properties never reached the reserve price.

As Campbell reflects, it demonstrated the ingenuity of farming communities to bide time for economic pressures to change. People forget banks were in the initial phases of staggering forced farm sales when Black Tuesday hit, triggering a rethink of economic strategy and forcing banks into the progressive phase. Then, the sharemarket saved New Zealand farming but with no alternative land uses, Landcorp (Pāmu), the state-owned farm, was press-ganged into buying run-down properties to keep land markets stable. Today there's carbon forestry.

The 80s equity crisis undermined the security of New Zealand farmers, creating new ways of production, business, financing and leadership. Farmers grouped to change things for themselves rather than wait for policy direction. However, insecurities shadowing farm financing and ownership, plus increasing industrialisation, colour New Zealand farming. Farms have become industrial assets at the expense of social and environmental resilience, thereby contributing to hardened attitudes toward alternative practices.

> This is where Warren Buffett's quote of farmers rolling an ostrich egg into the hen house and saying, "Ladies this is what the neighbours are doing!" comes home to roost – Branding Meat Quality

Eco-profit Observations

- Before the October 1987 sharemarket crash, more than 90% of non-performing and foreclosure bank portfolios were farm debt. Within a week it was less than 1%.

- Rogernomics collapsed land values as nobody, including banks, wanted farms because of low returns compared to the sugar rush of sharemarkets.
- The sharemarket crash separated the farm equity crisis into two phases, hostile and progressive.
- The hostile phase involved switching from assessing the farming family to only using land asset values as sole determinants of ROI and drew a line between farmers and their support network.
- Campbell identifies seven strategies rural communities used to beat back banker bullies: Entrenchment, women working often off-farm, black economies, creativity lifting of farm land values, hiding cashable assets, transferring lending to other sources, orchestrated public events and activities.
- Insecurities shadow farm financing and ownership, plus increasing industrialisation, colour New Zealand farming operations and hardening attitudes towards alternative practices.

42. Mark Warren - Basic Business Lessons

In a lifetime of pushing limits, Mark Warren suggests if you want to know the true value of someone's product, ask their opposition's opinion. If good, you'll know the product and have found a trusted adviser.

Mark owns Waipari Station, a 1,300ha coastal block in Hawke's Bay that's been in his family since the 1880s. He's a doer. Whether as a six-year-old collecting refundable beer bottles from parks and teen party spots at 5.30am on Saturday mornings or as a student adding value to grain by selling pork to ski club members, his seizing of business opportunities and acting on them is well known. His journey is laid bare in an autobiography, Many a Muddy Morning.

The disastrous drought of 1984 shoehorned him into the manager's role. Waipari is steep and unforgiving Class 6 land. Calving and lambing were 60% and 56% respectively. With working expenses at 99% of gross farm income and all staff gone, he started from scratch with overseer Mark Williams. What is clear from Mark's journey is having a business head drove him to challenge traditions. By the time he'd won the 1992 Hawke's Bay Farmer of the Year title he'd realised a return on capital of 15.9%.

This was during a time of deep rural crisis. Investors pumped money into sharemarkets, not farms. Equity evaporated for many farmers. A nearby couple comparing returns from shares versus livestock found shares always had an advantage till the 87 crash. Then their share values evaporated.

That's an important business lesson. Mark's accountant Pita Alexander will tell you, return *of* capital is more important than return *on* capital. As Mark observes, experience arrives in the split-second after you need it. Something health and safety auditors never appreciate.

So how did he challenge traditions? Here are three savvy business lessons for those in tight situations.

Firstly, focus on basics. Releasing staff allowed him to start with a clean slate, no egos. It helped him prioritise and price labour. Here another Alexanderism is useful: Be ruthless with money but gracious with people.

Then fencing. He hired a fencer first for a job that took years. Boundary fences came before internals. Once perimeters were stock proof, he could then chop blocks in half. Mark swung one new gate after work every day for the first month.

Secondly, employ nature. It's cheap and brings a lot of experience.

In 1983 Waipari carried 8,000 ewes squeezing out 4,800 lambs, a hangover from the Livestock Incentive Scheme. Reducing to 6,000 ewes producing 5,000 lambs was slightly better. It demonstrated how reducing numbers and feeding stock lifted production. Clearly, with all the fencing problems, overgrazing at low stock densities had driven down production. Today 4,000 ewes can deliver 6,000 lambs.

Mark noticed lamb survival was 5% higher when tailing hill blocks than easier country. Dropping lambing beats eliminated seasonal labour and strengthened the ewe flock through natural culling. If ewes get into trouble, Mother Nature makes sure it only happens once. Hardier sheep save labour and inputs, improving efficiencies.

Another efficiency was dropping summer lamb shearing and carrying their fine wool tip into winter. This dipped their micron inside 33.4 gaining premiums with a heavier fleece. Carrying a summer fleece also reduced flystrike because fly prevention chemicals need grease to improve their effectiveness.

> Furthermore, by flushing their best beef cows, they're not limiting each one to raise a single calf each year. That lifts herd genetic gain and production statistics faster – Tom Hargreaves.

His coastal country grows more winter grass than further inland. By dropping cow numbers from 500 to 300, he lifted cow condition and grazed 250 calves through winter to cash in on spring premiums. This eliminated hay and its costs.

He also left 100 bull calves entire because Angus crosses were hardier and a better fit for his country than Friesian crosses. Their 10% higher growth rate and 10 cents/kg higher premium yielded 3% more generating an additional $1,500 from the herd for free. After several years selecting bulls from his calves, he found their progeny differed little to stud purchases. Another cost he wound back.

Another way to let nature do more is forestry. It increases returns on marginal country barely running 5hd/ha, while reducing hill slippage and soil erosion. A 50c pine seedling in 1987 becomes a $150 tree 25 years later with few expenses beyond sunlight, CO_2, water and time. Essentially, forestry allowed Mark to pay-off Waipari within his working career as well as creating options for carbon credits and family succession. A quarter of the property is in trees covering steep hillsides without dropping any livestock production.

Thirdly, gain as much as possible from beyond the farm gate.

It's said 80% of New Zealand's lamb crop is unmarketable because it lacks consistency. Mark bought the district's first electronic scales. This allowed him to gain insights into lamb weights, giving him an upper hand and keeping livestock agents honest.

If farmers hold lambs back for heavier weights when premiums drop, they risk the whole market because buyers start looking elsewhere – Bill and Sue Garland.

Then, being invited into the Atkins Ranch American lamb-marketing scheme in the early 1990s was fortuitous. While originally growing hogget wethers for wool, he landed 15-20% premiums for 18-24kg cwt spring lambs. Today Atkins pays farmers to do their own drafting and for keeping to a production schedule. This is an important industry lesson, rewarding farmers for their preparation so the whole supply chain benefits.

Mark also built a business around his love of 4WDs, developing Hillseekers 4WD NZ, a specialist 4WD advisory company. That brought him into contact with industries beyond rural New Zealand to benefit from their insights and ideas.

So, what's the overall business lesson here? Work to accumulate benefits over time. Many basic business principles allow farmers to capitalise what nature provides for free.

Eco-profit Observations

- Return *of* capital is more important than return *on* capital.
- Reducing stock numbers and feeding stock better lifted production.
- Lamb survival was 5% higher on tailing hill blocks than easier country. Hardier sheep save labour and inputs, improving efficiencies.
- Dropping summer lamb shearing and carrying their fine wool tip into winter increased weight and dipped micron low enough to gain a premium.

- Without summer shearing, more fleece grease improves the effectiveness of fly-prevention chemicals.
- Grazing coastal winter country reduced hay and its costs.
- Leaving bulls entire cashed in on better growth rates and premiums.
- Selecting his own bull calves, finding after several years their progeny differed little to purchased bulls.
- Forestry creates best returns on marginal land often without dropping livestock production.
- Weighing livestock created more consistency and kept livestock agents honest.
- Pay farmers for product preparation and the whole supply chain benefits.

43. Geoff and Barbara Croker - Farming without Fertiliser

Stockman Grass Farmer editor Allan Nation observes large properties of better classes of land are valued much closer to their financial potential than smaller, poorer fertility properties. Farms with better fertility are easier to pay off through production.

However, New Zealand's most over-valued land is found in the high country, especially Central Otago. Hence the rush there into horticulture and high-end tourism enterprises to run down debt. But what options are there for low fertility, challenging properties in locations that offer fewer options? Can you farm without fertiliser?

Geoff and Barbara Croker found out the hard way you can. Their story is one of serendipity leading them down a different path.

Buying a Wairarapa hill country farm in the mid-1980s with an interest rate at 23% made Geoff and Barbara look at their options. They discovered very quickly they couldn't spend money on both fertiliser and fencing, so one had to go. They chose to eliminate fertiliser. Each year they waited to see if industry advice was correct and pasture production would collapse. Some 35 years later they're still waiting.

The financial case for this decision is simple. Fertiliser has a return on investment of weeks, fencing's is years. It means for the same sum of money, payback from fencing lasts much longer - some 80-100 years compared to fertiliser's few weeks. Nothing grows grass better than wire because controlling livestock is the cheapest form of precision agriculture. Even collar technology can't compete because it's a subscription.

> Michael realises now they were thinking the wrong way. Their new plan involved spraying crops, reseeding pastures and heavy fertilising – Michael and Catherine Reilly.

Fencing solves the greatest problem facing New Zealand farmers, which is fertility transferred from the middle of paddocks to shelterbelts, troughs, hay racks, gateways, hill tops and campsites. Increasing stock density and limiting where livestock congregate for long periods eliminates nutrient transfer. The easiest tool for that is fencing. The only additional expense is labour.

And here's something else worth thinking about. Removing maintenance fertiliser allows farmers to service more debt and grow their land holdings.

Depending on the property cycle, what if it's more cost-effective to buy additional land than investing that same money in fertiliser to maintain pasture performance of the home property? In other words, when would appreciating capital values of small, highly productive parcels of land outperform accumulating revenue from sustaining static production on home farms through maintenance fertiliser? Is that a pathway for family succession?

In the Crokers' case, dropping fertiliser created another opportunity.

During the early years they kept discovering ewes in excellent condition despite raising two cracker lambs on unfertilised hill country. Before long, they had 100 of these to start a small elite flock and they focused on building constitution as the basis for growth to weaning. This is how the Glenview Romney and then later, the South Suffolk, rams studs began.

Today, ewe hoggets are born on this country and then raised, weaned and mated, and lamb there, too. Only after weaning are these hoggets shipped off to another property as two-tooths. Having another property to hold ewes was a way of expanding their mob without buying more land.

Hoggets go to Matt Olds' farm for mating as two-tooths. Matt bought a summer-safe property at Mikimiki from the Crokers in 2018, but three years later another property at Tiraumea came up. So Matt sold up and with Geoff and Barb bought the larger property. All lambs born there are Matt's and the Crokers buy back the top 80% a bit above market value. This move allows the Crokers to expand flock size and helps a young farmer into the industry.

Lambing percentage is 145-165% while scanning is 190-200% with triplets included. As there is a big focus on triplets, all scanned triplets lamb at home to give those ewes more room. Growth rates for ewe lambs are 40kg liveweight at a late weaning of 120 days. That 40kg average weaning weight includes lambs out of hoggets.

Ewe hoggets are grown on low octane feed and have to perform at the highest level whereas ram lambs are moved to a finishing block and grown out on fertilised pastures. Research shows another advantage for growing hoggets on low octane feed is they do better during dry conditions compared to those raised on lush feed. This creates an unfair advantage because when livestock know what to forage in a drought, it saves farmers money through less feed wastage.

Ram lambs are sent to a block down the road at Taratahi. This block is fertilised with dicalcic phosphate every three years, challenging the need to fertilise every season. Over winter it carries half the ram lamb flock while the rest grow out at home with all grass wintering, no crops. Again, this keeps costs down and reduces environmental problems that come with winter crops.

Animal health issues are minimal because there's no problems with copper deficiencies or facial eczema. Ewes get rock salt before lambing and lambs are drenched when necessary. However, the animal health bill is high because it includes DNA testing to determine sire lines.

As Allan Nation notes, properties with high natural fertility tend to increase in value when recession hits, therefore it's often run-down properties that are on the market at this time because they have no buffer with stored fertility.

Increasing rainfall absorption and water-holding capacity is a priority. Dung beetles help this aim – Hamish Browne and Robyn Budd.

For those who buy run-down properties, the Crokers' example offers another path. By developing a breed that survives and performs in a low-input environment, they've demonstrated how genetics can eliminate the need for capital fertiliser.

Eco-profit Observations

- Smaller, low fertility properties seldom valued at real financial potential.
- Investing in fencing was savvier than investing in fertiliser because rate of return lasts longer.
- Pasture production hasn't collapsed without fertiliser over 35 years.
- Fertility shifting from the middle of paddocks to elsewhere perpetuates fertiliser applications.
- Removing maintenance fertiliser creates flexibility to buy more land.
- Identifying cracker ewes with twins on low fertility country started their ram studs.
- Hoggets raising lambs on low fertility country concentrated resilience and production genetics.
- Growing two-tooths out on another property saved buying more land to run more sheep and created income for a young farmer.

- Ewe lamb growth rates produce 40kg liveweight in 120 days on low fertility feed.
- Ram lambs are grown on fertilised pastures.
- Run-down, lower fertility properties have less buffering to climatic and market problems, hence tend to come on the market during recessions.

44. Graham Smith - Profiting from Paulownias

Graham Smith dairy farms near Wharepuhunga, North King Country. And Graham knows how to use nature for a profit. After buying the property, he bought 130 angora goats to clean up gorse, blackberry and broom to turn into fertiliser. They cost 50 cents each. He sheared them each year for three years and then sold them for $7.50. Point proven.

Before that he was sharemilking 420 cows, a very big herd in the 1980s. When he couldn't get a bank loan to buy 220 cows for one sharemilking job, the dairy company (Waikato Valley) stepped in and became his guarantor. In those days dairy companies would back the person, not the farm.

Unlike many of his mates, he went small. Now he still enjoys milking cows, with no staff and no burnout. To buy the farm, he had to run 100 cows even though it couldn't feed them. Now he milks 77. He has kept the same genetics for over 40 years so his herd's BW is 338 and PW is 391. Interestingly, the top cow family doesn't get mastitis as their udders don't leak.

He has been milking once-a-day for the past four years and enjoys it. The grazing rotation never drops below 22 days and pushes out to 60-70 days in summer. He plans to be at 36 days by December 10 to protect summer pasture. To ensure enough cover by January 1, he uses silage to push grass ahead of cows. If needed, he fells up to three trees per day to feed cows as well.

Final feedback from judges only focused on lifting monitoring efforts for future compliance, nothing practical such as specific trees to improve landscape function– Shane Armstrong.

He started planting trees after seven floods in two months in his first year. Back then, Environment Waikato paid farmers to plant trees along waterways and hills. In the early 1990s someone advertised for timber, so he planted paulownias. That started his forestry interests. He supplied timber to one guy but then decided to sell it himself. Now he's New Zealand's lone paulownia supplier with 2,000 trees, a third of his total forestry crop. He's possibly the only timber grower making money out of deciduous trees.

Globally, millions of paulownias are planted and harvested on a 10-to-18-year cycle. They coppice well, however Graham's unsure of longevity although he's milled an 80-year-old tree. They're great for heating but while their heat output is less than hard woods, they grow much more biomass. The trees are milled sustainably and because they coppice their numbers don't change, so carbon credits are unaffected. That allows Graham to claim credits in the permanent 50-year category and he's already accrued $64,000 over three years.□□

Paulownias are fussy. They need specific conditions to thrive; mostly no wind and free-draining and fertile soil. They need fencing from cows and calves as the trees are easily ringbarked, especially by cows. Bulls just rub against them. Fallen leaves are eaten immediately.□

> Steep hillside bananas could generate more income than grass, even from pups which emerge from underground corms – Ross and Karen Potter.

One hundred stems will drop 10 tonnes of sunflower-size leaves. Graham prefers a good frost in May, so all leaves drop at once for baling. One bale was so heavy he couldn't lift it with his Ford 6640 tractor. They are great in silvopasture grazing operations.□

Seeds are not easy to grow and are genetically diverse, which means it's easier starting with cloned tree cuttings or dug-up roots.□After 12 years paulownias send out a lateral tap root 50 metres and sprout new trees. Digging these up to put in a nursery is how he transfers root cuttings.□

Transplanted roots push up a trunk some six metres in the first year and snapping off permanent branches during this time creates the best clear wood, with only normal pruning needed after that. Frost destroys growing tips so maximising height in the first year is critical, then they grow from a spur. At this point they are dug up and planted elsewhere.□ □

Graham starts planting trees in May and June.□Rows are planted northeast to southwest. Earlier plantings were spaced 5m apart in 15m rows but are now at 20m x 20m to strengthen grass growth because paulownia leaves create plenty of shade. He's worked out that 100 stems/ha is too dense for solid pasture so he's dropped every second tree to let in more light at 50 stems/ha. Each tree is worth $1,500.

Graham also interplants rows with tree lucerne as its leaves are favoured by cows in summer. Paulownia is palatable because of fewer tannins in bark and leaves. Insects leave paulownias alone because their tannins penetrate the wood.

Overseas, trees are peeled and milled for musical instruments. Graham's biggest local client is a waka builder as paulownia is lighter than fibreglass. A fish lure maker uses shorter pieces and sells lures like colourful cigars into the US. Most buyers are replacing balsa wood as paulownia is stronger, more stable and doesn't twist. Graham also exports to a Singapore museum which uses the wood to repair 2,000-year-old paulownia artifacts from China.□

While the paulownia has some specific requirements, it's essentially no different from many other tree crops. Yet it has zero appeal to New Zealand agriculture.□□□

Graham's had many New Zealand scientists visit and check out paulownia's amazing array of properties, but, for whatever reason, nothing happens. "A $500,000 trial will provide all the proof if anyone wants to fund it," he says. "Paulownias are great at soaking up excess nutrients.□Studies from Louisiana State University show 100 stems/ha will soak up to 70% of excess pasture nutrients." Fallen leaves release these nutrients slowly allowing good uptake by pasture. Rotting leaves and branches enhance soil carbon levels.□

Yet lack of official interest surprises and frustrates. In a world looking for natural solutions, why do so few industry professionals appreciate insights from farmers like Graham Smith?□

Eco-profit Observations

- Bought goats to clean up woody weeds, sold them three years later for 15 times their cost.
- Staying small meant no staff or burnout. Still enjoys milking cows.
- Top cows don't get mastitis as udders don't leak.
- Grows trees to feed cows if dry.
- Seven floods in two months got him interested in trees.
- Paulownias very versatile, both for timber and livestock but cloning is easier than raising from seed.
- Best at 50 stem/ha for good grass growing in silvoculture operation.
- Paulownias replace balsa wood because of weight and durability.
- Planting Paulownias generates many environmental benefits.

45. Michael Magiera and Smita Apte - Question Every Expense

So, how does someone without a farming background end up milking 1,600 cows on 1,250ha in the Catlins? In 1994, Michael Magiera and Smita Apte sailed to New Zealand on other people's boats. While Smita completed a PhD on greenshell mussels and crayfish, Michael picked up dairy employment. They ended up in South Otago and in 2005 bought a farm at Owaka. Since then they've added eight more blocks to the business.

Coming from Cologne, Germany's pharmaceutical capital, Michael couldn't cope with its smell last time they visited, it was like breathing inside a chemical storage shed. Many farmers look for alternatives to get away from agrichemicals. While not organic, their focus is the economic farm surplus/kgMS (EFS). Nothing else matters.

Focusing on EFS got them questioning every expense. They bought their second property during 2009's GFC. With no capital to expand their cowshed they moved to once-a-day (OAD) milking. That practice eliminated all fertility treatments, meaning no more inducing non-cycling cows. They still do heat detection using an EZHeat camera but if cows aren't needing interventions they stop monitoring. They do not monitor for, nor intervene with, non-cycling cows. There's no need for electric collars as they don't have problems collars detect.

Dropping calf meal emerged from asking a neighbour who had milked cows in the 1960-70s how to reduce costs. He replied he'd never used it. Now they'll never go back. Calves are on pasture in 10-15 days and wean in 10-12 weeks. At a conference, Michael heard a scientist say grass nutrient composition was much closer to milk than calf meal. Gone are heavy bags, feeding birds and sick calves, although older calves do get other health problems such as yersinia.

Because they initially bought animals from all over the place, about 35 cows succumb annually to Johnes. Experience showed 73% of medium and low positive reactors on blood tests record as negative a year later. Reactor cows and their calves were culled and their milk not fed to replacements.

Those tests, plus lost calves, cows and production for the following season generated losses up to $200,000. Now any cow showing signs of Johnes is culled immediately as observation is cheaper and more reliable than testing. Michael wonders if stomach tubing every calf with colostrum made calves healthy initially but spread Johnes.

No dry cow or teat seal has been used since 2021. Michael recognised the futility of hanging on to high somatic cell count (SCC) cows because dry-cow therapy didn't fix them. Furthermore, the earlier such high SCC cows are culled, the more savings in feed costs. It also reduces SCC in later lactation and overall SCC in the following season. Culling the threshold for SCC dropped the count from 800,000 to 300,000 in the 2024/25 season.

They rear 60 Jersey bulls annually for R2 heifers and use natural mating after AI to be self-contained. Their bull lines only come from the top 20% BW cows. When sourcing semen for AI, they target negative SCC, positive high fertility, strong udders with outstanding udder support, high capacity and A2A2. Most of herd is A2A2 but they haven't gone for premiums as the bonus is less than 10 cents/kgMS.

The value of the large mob became apparent in spring. Pasture growth was strong and even across paddocks. Campsites and tag disappeared. Hillsides grew grass, leading to better land utilisation – Max Nicol.

SCC and milk urea nitrogen (MUN) rates mimic each other, with both being higher in spring but settled in October. They don't measure or follow MUN but have empty rates of 7-12% for cows and 3-9% for heifers.

They loved urea but when payouts dived in 2009, that got axed. Michael noticed the pasture was always better with more clover growth in corners or in areas missed by spreading. Dumping urea lowered pasture growth for a couple of seasons and coincided with clover root weevil arriving. Then they started introducing red clover.

They use ryegrass only on the flats but are now experimenting with cocksfoot and timothy (only Greenlea Cocksfoot) plus two white clovers, red clover, plantain and chicory. They started using cocksfoot on hills after buying a property with cocksfoot. They found it held longer into the dry and showed more persistence than ryegrass when hammered with wind, drought and snow.

Winter crops are incorporated into their pasture renovation plan. First, two turnip crops are planted, then kale for a second winter before going back into grass. They also grow swedes, plus one or two kale crops before regrassing.

Ryegrass domination reduced when changing to longer pasture recoveries to favour deeper rooting grasses such as prairie grass and timothy. Red clover appeared in paddocks it'd never been sown in over 20 years – Phil and Viv Gray.

Winter milk cows graze a kale crop and later calvers get a turnip crop in late August-early September at a ration of a third turnips, two-thirds silage back at the barn. They can't ditch winter crops as they assist in building spring covers while providing flexibility for dry autumns.

Again, looking to save money led to investing in solar panels. With OAD milking, the milking shed only operates in daylight hours, meaning no summer power bill and reduced costs for other months. This adds resilience in case of power disruptions with solar, battery and diesel generator backup.

To deal with wintering issues they built a composting barn but doubt its economic viability. It requires 2-4 hours tractor work daily but during extended damp, cold periods, the bedding pack loses heat and doesn't generate enough evaporation to dry out. That requires working over with tractors until it generates heat again. Working in two tonnes of lime every 2-3 weeks has gone well for the last two seasons but they still have problems.

So, what are important business lessons here? Observation is cheaper than technology and removing an input is better than minimising its use. This is not industry policy. An example is Fonterra's low GHG footprint per kgMS initiative. Michael calculates it'll cost 20 times more than the bonuses paid to achieve production efficiencies. This is the difference between theory and reality.

Eco-profit Observations

- Once-a-day milking eliminated fertility treatments for mating.
- No electric collars as they don't have the problems collars detect.
- Grass nutrient composition was much closer to milk than calf meal. Calves on pasture in 10-15 days and wean in 10-12 weeks.
- Any cow showing signs of Johnes is culled immediately as observation is cheaper and more reliable than testing.
- No dry cow or teat seal since 2021. The earlier such high SCC cows are culled, the more savings in feed costs while also reducing SCC in later lactation and overall SCC in the following season.

- Breed own bulls from top 20% BW cows. Genetics target negative SCC, positive high fertility, strong udders with outstanding udder support, high capacity and A2A2.
- Noticed pasture was always better with more clover growth in corners or areas where urea missed. Dumping urea lowered pasture growth for a couple of seasons and coincided with clover root weevil arriving. Started from there with red clover.
- Find cocksfoot holds longer into the dry and shows more persistence than ryegrass when hammered with wind, drought and snow.
- Can't ditch winter crops as they assist in building spring covers while providing flexibility for dry autumns.
- Adding solar power eliminated summer power bill, reduced costs for other months and added resilience in case of power disruptions.
- Built a composting barn but doubt its economic viability.
- Meeting Fonterra's low GHG footprint per kgMS initiative will cost 20 times more than bonuses paid to achieve production efficiencies.

46. Debt Creates Futures

Only women understand profit. Men are too emotional.
Grahame Rees, KLR Marketing. Bathurst, Australia.

Most problems with farming do not originate at ground level but six feet above says Pita Alexander, founder of Christchurch accounting firm Alexanders. He has spent 60 years helping farmers understand what makes good business. He observes the best thing farmers can do for their children is love their mother. After spouse and children, debt should be next, he argues, especially if enterprises driving farm businesses are wobbling.

Debt is often essential to get started. There's good debt, bad debt and ugly debt. Good debt involves tax-deductible interest payments secured to an asset that's somewhat inflation proof or able to replace itself. This is the debt you want. As Pita says debt leads to progress, scale, viability, profitability, job satisfaction, and something to leave your family. Owning is best, after all, nobody washes a rental car.

Bad Debt occurs when paying interest rates with after tax cash. Much residential debt fits here, what Pita calls life support debt. Really ugly debt combines non-tax deductible interest payments with climbing interest rates being secured to a non-appreciating asset. This is treadmill borrowing and if associated with unlimited guarantees to banks, owners can be over-exposed even with small loans.

Why reduce debt? Pita observes there's not much else that delivers an 8% return with cash surpluses. Debt reduction is compounding as it reduces payments for this and every other year thereafter. It increases job satisfaction by accelerating confidence and accessing better future options. It opens doors to create financial reserves to invest within or beyond the business. The greater your financial security the more choices you have. His best clients know their profit at the start of the season, not aimlessly waiting for what is left over. Often, making choices before profits materialise leads to disasters.

The path to spiralling debt is easy. Never forget the law of 72. Dividing interest rates into 72 determines how long an investment or debt will double in value. For example, $100,000 at 8% will take nine years to reach $200,000, at 12% it'll take six. At 3% inflation, the cap most developed country's reserve banks aim for, prices double every 24 years but in reality, can take less time because of global black swan events.

Examples of Misery Index

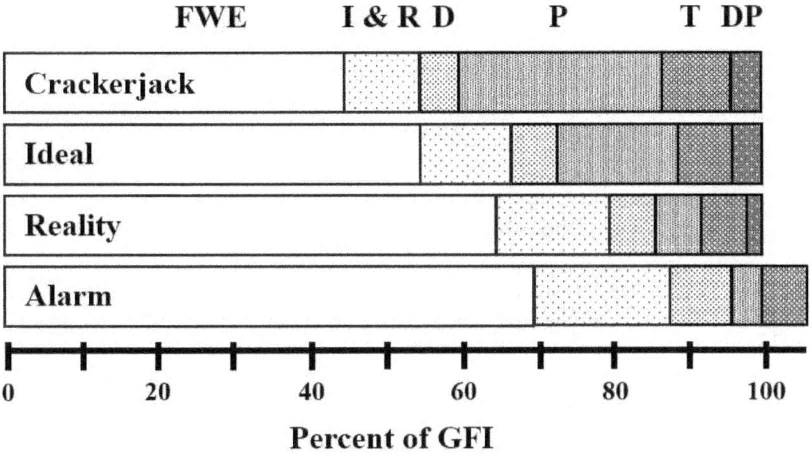

Misery Index Examples	Cracker	Ideal	Reality	Alarm
FWE – Farm Working Expenses	45%	55%	65%	70%
I & R – Interest and Rents	10%	12%	15%	18%
Misery Index	**55%**	**67%**	**80%**	**88%**
D – Depreciation	5%	6%	6%	8%
P – Personal drawings, etc	27%	16%	6%	4%
T – Taxes	9%	7%	6%	6%
DP – Debt Principle	4%	4%	2%	0%

One tool Pita developed to indicate farm business health is the Misery Index. It reveals how strong a business' profitability is and at a higher level can indicate the industry's pulse. It's the sum of:

1. Farm Working Expenses (FWE) as a % of Gross Farm Income (GFI) Ratio; the proportion of income sunk in tax deductible costs to run an operation. Ideally this should not exceed 50% GFI,

2. Interest and/or rent paid as a % of Gross Farm Income Ratio. This represents access to land or other productive assets, the foundation of farm businesses. Ideally this should not be over 10% of GFI and be disturbed if closing in on 30%.

If the Misery Index consistently exceeds 75% of GFI, Pita realised cash surpluses struggle to generate financial reserves. When over 95%, expect the farm sold within five years. Ideally getting this figure below 65% is best to produce real cash surpluses, which preferably, should be above 33% of GFI. Many couples never get above 20%.

A true cash surplus must cover depreciation (5%-7%), personal drawings, superannuation and allowances (6%-7.5%), income taxes (6%-7%), and debt principle (2%-4%). If depreciation is running 7% and Misery Index is over 80%, doesn't leave much to cover everything else. There are crackerjacks that work at successfully breaking one benchmark, but never two.

This table from Alexanders' client data illustrates average misery indexes across five classic farm business categories ending 30th June 1998. Note, hill country sheep and beef properties are above ideal and mixed cropping and dairy are on the cusp; alarm bells are ringing. Over the eighties and nineties, many of Alexanders' clients were achieving 70-73%.

Examples of New Zealand Farm Types and Misery Index and Tombstone Ratios

Farm type	FWE % GFI	Interest/rent paid % GFI	**Misery Index**	Average net profit $	GFI to plant ownership ratio	Ideal Tombstone Ratio
Dairy owner operator	54.1	20.1	74.2	53,823	3.7	4.5
50/50 sharemilker (no land ownership)	53.4	15	68.4	65,143	5.9	6.0
Mixed cropping	61	14	75	63,986	2.6	2.75
Sheep & Beef downlands and flat	53	14	67	35,439	3.6	4.0
Sheep and beef hill country	56	24	80	25,303	4.2	4.5

By early 2000s, many were below 70%, most likely reflecting income flows from a low NZ dollar. Today (2024) FWE is 10% higher resulting in Misery Indexes of 85%. In the last 50 years, New Zealand annual farm income has only outpaced annual farm costs twice.

Research and consultants constantly mention it's not types of technology, crops, livestock or whatever else farmers use which determines profitability, its financial discipline as well as the ability to observe surroundings and act which matters – GMO Ryegrass only Treats Symptoms

Another financial indicator measures iron disease. Pita believes 10% of New Zealand farmers suffer from this.

If GFI is twice the book value of machinery and plant (2nd column from far right), it's due to not enough income, over capitalisation with machinery or a combination of the two. The ratios for sustainable profit and machinery/plant efficiency are in the far right column. Note all are less than ideal.

Stockman Grass Farmer editor Allan Nation would've called this the Tombstone Ratio after blue silage silos standing beside every American dairy barn. Lower numbers mean more invested in depreciating capital and exposure to inflating energy costs. Selling that old spare tractor worth $1500 could buy several calves that appreciate over 20% inside 12 months. Leasing equipment can indicate a low Tombstone Ratio and not enough business profit for affording new gear and depreciation rates.

If families have strong farm equity while making low profits, often its job satisfaction keeping them there. What gets farming families into trouble are chasing unrealistic expectations, procrastination and inabilities to prioritise available cash, resources and time.

Resilience is like a red meat muscle - it needs to be worked to get stronger – Why Bear Grylls should Spearhead Red Meat.

Many a wise word whistles through false teeth. Pita observes attitude is much more important in times of stress than deficits. It encourages open communication and listening; women are better at putting issues clearly on the table and children's opinions are seldom tainted by farming baggage. Farming families face more down-cycles than up-cycles. So, if at first you don't succeed, boldly ask if the loser gets anything.

Eco-profit Observations

- Most farming problems originate six feet above ground level.
- Good debt involves tax-deductible interest payments secured to an asset that's somewhat inflation proof or able to replace itself.
- Really ugly debt combines non-tax deductible interest payments with climbing interest rates, secured to a non-appreciating asset.
- Debt reduction is compounding as it reduces payments for this and every other year thereafter.
- Dividing interest rates into 72 determines how long an investment or debt will double in value.
- If the Misery Index consistently exceeds 75% of GFI, cash surpluses struggle to generate financial reserves.
- In the last 50 years, New Zealand annual farm income has only outpaced annual farm costs twice.
- If GFI is twice the book value of machinery and plant it's due to low income, over capitalisation with machinery or combination of both.
- Lower numbers mean more invested in depreciating capital and exposure to inflating energy costs.

- What gets farming families into trouble are chasing unrealistic expectations, procrastination and inabilities to prioritise available cash, resources and time.
- If at first you don't succeed, boldly ask if the loser gets anything.

Adaptation: Debt Creates Futures. In Practice #224 November/December 2025. Holistic Management International.

47. Roger Slee - Once-a-Day Winner

Roger Slee saw New Zealand's sheep industry demise coming. He recalls a pre-season Waitaki Farmers Freezing Company meeting in 1986 where he challenged CEO and chairman Athol Hutton about surviving on $12 for a 12-14kg cwt lamb. Hutton shouted at him he was 'bloody lucky farmers were being offered that!' Within six weeks there wasn't a sheep on his Kurow property.

The Southlander started farming in 1964, aged 16, with his 21-year-old brother Syd purchasing Te Mara, 1,400 acres in Western Southland for NZ£33,000 with a deposit of 12% and paying 5.5% interest to the State Advances Corporation. With a tractor borrowed from their father they started developing the property and poured every dollar back to lift flock numbers from 1,100 ewes plus replacements to 5,000 in six years.

In 1972 they bought Charnwood, 11,000 acres in South Canterbury. Roger managed the development from Te Mara partnership's income. However, it was costly to run and in 1981, with cheap fertiliser and other subsidies looking wobbly, they sold. By then Roger had bought 640 acres at Kurow on his own account with good soils, reliable climate and border dyke irrigation to manage the dry.

With money tied up in his brother's operation and loans for his own, Roger was highly geared so needed more income than 4,000 Coopworth ewes with 125% lambing provided. With interest rates climbing and sheep prices sliding, he began looking at pipfruit, stone fruit, seed and cereal crops.

With banks looking for figures that suited their growth models, and the Berryman farm taking a different approach to sustainable farming, they did not feel confident finding other like-minded business partners – Peter Berryman.

But there was one overbearing issue; none of these industries had a single professional marketing body to export produce. Most horticulturalists viewed their industry leaders' abilities as suspiciously as he did the sheep industry's. At this time, meat export companies competed against each other. Dairy was different, as all exports went through the NZ Dairy Board. Then Temuka Dairy

Company, 100 miles away, said it would come south to Kurow for 300 cows, a large herd back then.

All the locals said he had rocks in his head and wouldn't join him, but he went ahead and put half the farm into dairy, starting a three-year contract with a 50/50 sharemilker. That meant he didn't have to buy livestock and could look over the contractor's shoulder to gain experience. His was the eighth dairy farm on the Waitaki plain, a decade before the South Island dairy boom.

After the exchange with Athol Hutton, Roger started milking on his own with 500 cows in a 36-a-side herringbone shed. He managed to buy land from neighbours, built a 60-bale rotary shed and milked 1,000 cows on 900 acres. It was a very profitable business. Initially getting money to expand was challenging but once he was milking 1,000 cows, banks were falling over themselves for business.

As a sheep farmer, he thought dairying would be great if milking once-a-day. The mid-1990s saw him in a joint venture with a staff member who bought a property next to his own father's farm to combine both for once-a-day milking with 400 Jerseys. They did twice-a-day for two years to benchmark production. In year three a quarter of the herd went once-a-day using lower-producing cows. Results from herd testing the entire herd eight times encouraged them to milk 50% the next season. Again, milk production didn't drop so in year three all cows went twice-a-day till three weeks before mating, then once-a-day.

For example, humates with urea lift pasture production by 20-30%, mostly from reducing nitrogen leaching by 60% - Jeff Broomfield.

Now Roger says he would never go back to milking twice daily. Management firm Farm Right's entire team visited to compare actuals with a budget assuming twice-a-day. They proved Roger was $30,000 better off than their calculations predicted. The cash surplus came from reduced labour, animal health and supplements; no PKE, grain or maize. Any surplus pasture was made into balage or silage. Calving spread was also tight, there were fewer lame cows and cows lasted an extra two years. It all added up.

He thinks industry fears of lower production stifled enthusiasm for once-a-day, although he proved the opposite. Maintaining pasture quality is more challenging milking once-a-day but Roger topped pastures to keep quality up.

Furthermore, alongside the 400-cow operation, Roger pushed his 1,000-cow farm to 1,500 with the arrival of an enthusiastic contract milker. Going totally cut-and-carry from two leased farms five and eight kilometres away wore out tractors and feeding equipment and required a team of rostered staff to milk twice-a-day. Yet the 400-cow dairy made more money per hectare and per cow. The 1,500-cow venture went for three years and dropping back to 1,000 cows reduced stress for both himself and the contractor. He doubts it ever paid for itself.

This was 2003 and while dealing with cancer he bought another property with another business partner for a second once-a-day operation milking 400 Jersey cows. That worked well for 10 years. Keeping to an all-grass system, production reached 440kgMS/cow. They consistently had best in-calf rates and a tight calving spread despite never using interventions like cidrs. Vet visits were few.

When asked what happened between the ages of 0-15 to ground his success, he recalls working on a farm where the family, working together all the time, only discussed farming matters once a month at a formal meeting. However, his family had those meetings with every cup of tea, lunch and evenings. They were constantly reviewing their observations and challenging each other on management and ideas.

Frequent, passionate conversations engaging a sense of purpose appear to be the essence in detaching from traditions to grasp and seize opportunities. Those social skills may also be a reason why Roger had so many successful farming joint ventures.

Eco-profit Observations

- Foresaw demise of sheep industry.
- A single marketing entity got him looking at dairy.
- Started with a sharemilker to lower cost, spread risk and learn the trade, then went on his own.
- Once at scale much easier to get money from banks.
- Moving to once-a-day was a three-year process. Note, he started out with lowest producing cows to reduce risk.
- Reduced labour, animal health and supplements; no PKE, grain or maize produced a $30,000 cash surplus that couldn't be accounted for by management firm Farm Right.
- Tight calving spread, fewer lame cows and cows lasting another two years all lifted profits.
- Pasture quality more challenging on once-a-day.

- Four hundred cows once-a-day made more profit per hectare and per head than 1,500 cows on twice-a-day.
- Created an all-grass system that produced 440kgMS/cow, never used interventions, no vets.
- Frequent, passionate conversations engaging a sense of purpose can reduce attachment from traditions to grasp and seize opportunities.

48. Tony and Angela Blunt - Old School Financing

What's the cheapest way to buy land? Cash.

Tony Blunt did his father's books for 30 years and learnt how the old ways of saving cash reduced risk when buying farms. He realised that buying land with after-tax dollars resulted in lower payments of principal, interest and future tax and this quickly secured the freehold. The 1991 Young Farmer of the Year winner and current Kaikoura District Councillor is the proof that being frugal and savvy pays off.

Tony and Angela run an 800ha (450ha effective) coastal hill country property near Kaikoura. They're third generation but the first to purchase the whole property by saving up to buy in chunks over several years with cash, before eventually gaining a mortgage.

Compare that to what the industry prefers. A local bank offered a loan 25 times their current mortgage to buy neighbouring land consisting a third in river stones that drowned in winter and baked dry in summer. Tony calculated that area (15% of their entire operation) had to generate more cash than from the original property just to cover interest payments. The only thing for certain was they were a day closer to another downturn as farm values dropped. Bankers had just moved to commissions and the GFC happened a year later.

The home farm never generated enough profit to buy it in a timely manner, so farm purchases and practices must be thrifty and practical to free up cash. He saved $100,000 over three years shearing his own sheep.

Now debt doesn't affect his sleep anymore. While his income fell 25% in the 2023/24 season, farm working expenses fell 50% – Willie White.

Tony ran a fencing business for many years targeting Banks Peninsula's most difficult jobs. That meant leaving at 3am, driving four hours to a site, spending three 14-hour days on the job, then picking up a ute-load of cheap groceries in Christchurch on the way back. His Young Farmer of the Year award was an unfair advantage allowing him to land some contracts while saving cash in the supermarket.

Another passion for Tony is wool, especially Drysdale, the best carpet wool. As wool prices crashed in the late 2000s, he joined with other Drysdale growers to restrict supply to keep prices above $7.00/kg. When all their efforts were shut down in 2015, 20 years of beloved breeding was sent to the abattoir in a single afternoon. Drysdale lambs have the highest survival rating because they are born with an inch of wool, more than a match for Antarctic southerlies.

A decade on, he's winding down his Romney flock and increasing Halfbreds for better wool returns. Crossbreds eat too much for the income they generate, especially with current wool prices. Tony calculates two Halfbreds with 110% lambing will beat a Romney with 150% lambing.

Drought is common on Kaikoura's coast but he's never sold capital stock or bought feed. Selling stock reduces herd memory for events such as snow. Cows graze up to 1,000 metres above sea level on rough blocks all winter and calve there. By running just over 7su/ha, he cashes in on green summers by making and storing balage to sell to dairy farmers when the dry returns and prices lift. Tony does the opposite of everyone else; buys when they sell, sells when they buy.

Somehow a John Deere magazine ended up in his mailbox early 2000s. Its article on soil carbon came from the Australian CSIRO. Drought was visiting again and he calculated lifting soil carbon by 1% annually would increase water-holding by 100mm over four years, allowing grass to grow for another month. That changed how he looked at grazing.

Now pasture grows longer during spring and summer (1,500-3,000mm) and then is trampled later for compost. He mows tag in autumn if need be. When sheep and cattle don't eat it, earthworms and soil biology will. Nothing is ever wasted.

Moving to all-grass significantly dropped costs and crop risk. When wet there's no strip grazing, even on pasture. Sticky clay soils pug easily under winter crops and bake hard in summer. He plants chicory, plantain, red clover, white clover, any tap-rooted species with cocksfoot and italian ryegrass and seaweed soil conditioners to encourage root hair growth.

> If pasture renewal is needed, he mainly sows chicory, plantain, red and white clover. He doesn't sow any grasses as the seed bank is already there – Andy

New ryegrasses don't last so he keeps the older varieties. This means he breeds stock immune to ryegrass staggers, as he did with Drysdales; the focus is fertility and survival. That resulted in lamb losses from scanning to tailing in single figures. It'll take many years to create the right sheep, so it is something his boys can take over.

Surprisingly, gaining consents for nine turkey nest dams is easier than one large one. Also, they'll be built closer to irrigated paddocks, meaning less pumping and pipe. He plans to switch from diesel to solar and pump all day drawing 10 litre/sec from river and to gravity-irrigate at night when there's less evapotranspiration.

In 2016, just before the big earthquake hit, Tony and Angela leased two PurePods sites, essentially two glass-clad cabins, each with a separate private view. As Tony observes, "Two are better than one as the cleaner is already out there." With a million people passing through Kaikoura annually, they jumped in early, and the initial investment was repaid within days, helping builders. Their profit-sharing agreement created cashflow utilising natural views of the farm, while they knew earthquake insurance was likely to take years to settle.

With his many community hats, Tony is sometimes bemused many farmers think they can only spend their way out of debt. Tony has spent a lifetime proving it's not maximising production that gets you ahead. It's looking for an unfair advantage from your personal abilities and your property's features.

Eco-profit Observations

- Buying land with after-tax dollars resulted in lower principal, interest and future tax payments to quickly secure freehold.
- Shearing his own sheep over three years saved $100,000.
- Drysdale lambs have the highest survival rating because they are born with an inch of wool.
- Winning a national award has allowed him to land some fencing contracts while saving cash in the supermarket on the trip home.
- Crossbreds eat too much for the income they generate. Two Halfbreds with 110% lambing will beat a Romney with 150% lambing.
- Selling stock reduces herd memory for events such as snow.
- He cashes in on green summers by making and storing balage to sell to dairy farmers when the dry returns and prices lift.
- Lifting soil carbon by 1% annually will increase water-holding by 100mm over four years allowing grass to grow for another month.

- Moving to all-grass wintering dropped risk and costs as sticky clay soils pug easily.
- Seaweed soil conditioners encourage root hair growth.
- Breeds stock immune to ryegrass staggers to use older varieties of perennial ryegrass.
- With irrigation, plans to switch from diesel to solar to pump all day, drawing 10 litre/sec from river and gravity irrigate at night when less evapotranspiration.
- Got into agritourism to cash in on the natural views of the farm.

49. Adrian Frei - Low-Cost Champion

"I don't want more than 400kgMS/cow as production has little relationship to profit," says Adrian Frei proudly. He wants consistent profits from lower costs through fewer replacements, low inputs and from technologies that reduce fossil fuels and increase labour efficiencies.

After migrating from Switzerland in 1995, he made his way south from Waikato to end up at Riversdale, Southland. "New Zealand offers the best opportunities to run efficient dairying," he says, and Southland with 200-240-hectare blocks has more scope than most. But a stint sharemilking for management firm Farm Right put him off intensive pastoral dairying with high stocking rates, no supplements and only nitrogen support.

In 2008 he started looking for a different strategy - without urea, superphosphate, potassium chloride or agrichemical sprays. He started down the biological path after meeting Steve Haswell from BioAg at the North Island Field Days. He was looking for a product to remove effluent pond crust and told Steve BioAg had no reps in Southland and that he, Adrian, was a bloody good bugger to work with. Two weeks later the phone rang - he had won $2,000 worth of products.

Figuring out how to work with nature has become a passion as his experiences reveal many agrichemical applications in conventional systems are uneconomic – James Halford.

By then he'd settled on 350 cows. To reduce production intensity, he bought more land, juggled young stock off-farm, leased land, changed equity partners and started heading into organics. Now (October 2023) a $2.00/kgMS organic milk premium makes up for lower production and an additional 20 cents for A2A2 is welcome.

Organic premiums mean his 120,000kgMS is equivalent to 150,000 conventional, his highest previous production. Now he's milking 300 cows twice-daily at a stocking rate of 1.6hd/ha and peaking at 2.3kgMS/cow/day. He reckons he grows only 8-9 tonne DM/yr, so while he is not a high pasture performer and while DairyNZ says cows need 12.7kgDM to produce one kgMS, Adrian has done it with 9.8kgDM. This is a calculation by DairyNZ when he was in the monitor farm programme for four years. Adrian puts that figure down to a shorter sward with more nutrients and slightly higher DM.

Moving to an all-grass system results in no crop and regrassing costs but some feed is wasted over summer. He thinks this saves $1,000/ha. Most paddocks are sown in 17 species of pasture but maintaining diversity is much more challenging. He's learning to bring back red clover by lengthening recoveries out to 60-70 days after Christmas. Multi-species pasture reduces boom/bust grass growth and milk production common with perennial ryegrass and the need for silage.

Another cost-saving is from the dual-purpose Fleckvieh breed, which provides good fertility and longevity, and holds condition better than Friesian, meaning they probably use less feed in winter. They milk as well as Friesians so need some sugar, but hopefully not as much molasses as Friesians.

He had a shocking 16% empty rate this calving, well over the usual 6%. The empty rate influences heifer replacements. A low conception rate was due to higher post-grazing residuals reducing available energy over mating. He noticed by week four grass had lost its energy because round length was too long.

During calving last season his herd suffered very little mastitis, no retained placentas and needed no antibiotics. Adrian puts this down to feeding good quality hay.

As a Swiss, Adrian knows cows are not fed silage when producing hard cheeses like Emmentaler and Gruyère because of the potential for butyric acid to compromise cheese integrity during curing. In looking to reduce fermented supplements he found hay making and drying technologies in Europe. He calculates that the latest hay-making regime using smaller tractors lowers contractor costs by 60%.

Instead of making silage from surplus spring pasture, he cuts and dries it. Hay is usually baled at 88-90% DM, but if left curing too long in the paddock it loses all the fine leaf and nutrition. He now bales at 65-70% DM and puts up to 60 bales in drying cradles. An 18-kilowatt blower pushes high-pressure air heated at 44C from waste oil to create a 90% DM bale of 11MjME hay. Hay is pH neutral whereas silage is acidic. While not anti-silage, any acidic herbages require more supplements such as salt.

Lower horsepower hay-making equipment includes a nine-metre-wide sickle-bar mower, requiring an 80hp tractor running at 1,360 revs doing 8km/hr resulting in 7ha/hr. A workshop robot sharpens blades after every 30 hectares. The hayrake machine is front-mounted so tractors don't run over feed while travelling at 14km/hr and an extra bonus is it has only one grease nipple. With two cradles, the maximum number of bales can't pass 60.

A large expense on many farms is synthetic fertiliser. Instead, Adrian is focusing on composting and recycling nutrients. Now his largest expense is sawdust and straw, all of which is composted. On top of that Adrian applies 15 litres of fish and minerals copper, zinc and selenium from several suppliers, including BioAg. He calculates his fertilising costs at around $80/ha for all bought-in minerals and conditioners; plus another $100/ha for sawdust and straw, all of which is composted and used on-farm.

> Peter came back with an idea - put the hardest part of the farm to irrigate into organics for five years, the rest would stay as it was – Tim Chamberlain.

To do this he built a composting barn with a base depth of 600mm sawdust and provides 9m^2/cow. Sawdust is better for composting than chips or shavings and he runs an aerator through it every winter day. In late spring, bedding is heaped and left to heat up with compost turning 5-6 times over summer once its temperature has reached 55C.

A bonus in achieving all these changes is an ability to operate without an overdraft and its interest payments. As his overall profit strategy comes together, Adrian's mental health benefits from that as well.

Eco-profit Observations

- Organic premiums mean 120,000kgMS is equivalent to 150,000kgMS with conventional price.
- DairyNZ says cows need 12.7kgDM to produce one kgMS, Adrian has done it with 9.8kgDM meaning better feed efficiency.
- An all-grass system with no crop or pasture renovation saves $1,000/ha.
- Is bringing back red clover by lengthening recoveries out to 60-70 days after Christmas.
- Multi-species pasture reduces boom/bust grass growth and milk production common with perennial ryegrass and the need for silage.
- He runs Fleckvieh cows for good fertility and longevity. The breed holds condition better than Friesian so he possibly uses less winter feed.
- Lower conception rate was due to higher post-grazing residuals reducing available energy over mating.

- Feeding good quality hay results in very little mastitis, no retained placentas and last season no antibiotics.
- Butyric acid compromises cheese integrity required for longer curing.
- Hay-making equipment requiring smaller tractors and lowers contractor costs by 60%.
- Can bale hay at 65-70% DM, rather than 88-90% DM.
- Fertilising costs at around $80/ha for all bought-in minerals and conditioners plus $100/ha for straw and sawdust.
- All composted fertility returned to the farm.
- Accommodates all these changes without operating an overdraft.

50. Daniel Wheeler and Donna Hicks - Farm Without Capital

Getting started in farming is an incredible challenge but Daniel Wheeler managed to find a way. As a Christchurch kid raised by a builder, there was no short cut into the industry. His journey demonstrates two things, how to get scale without capital, secondly how to use genetics to capitalise on what nature provides for free.

A stint at Massey University introduced him to genetics and his whole focus shifted to profiting from weaning weight. Meeting geneticist David Scobie attracted him to low-input breeds. New Zealand feral breeds offered hardiness insights. Mohaka were little, highly fertile sheep which were often lame during spring but came right as feed hardened. Campbell Island hoggets needed stress for wool to fall off to get pregnant. Too much wool and they remained barren for their entire lives. Daniel reckons they tasted the best.

Scobie used Daniel's Finn rams to introduce short tails into his AgResearch mobs. Finn is the basis of Daniel's low-input flock.

He once bought 100 Finn ewes from Grasslands research unit at Massey University only to realise he had no place to graze them. Luckily that gift appeared as his ewes were crossing Cook Strait. That evening started an ongoing adventure in securing nine blocks of land in and around Christchurch to double the area of his 36ha farm near Amberley. While most blocks have peppercorn leases, distance compromises grazing management. However, it has allowed him to grow a flock of 220 ewes and 380 lambs at home while trading livestock across other properties.

Daniel's main income is from pregnancy scanning and when not in New Zealand you'll find him in the UK. Therefore, leased blocks need to be small enough for fully employed partner Donna to check during the weekends Daniel is away. This means being picky with leases. Their three blocks on the road to Rangiora are easily checked there and back. They avoid properties where houses are at entrances so they can respect landlords' privacy on weekends and early mornings.

His desire is to produce the ideal productive and fertile sheep. By productive, it requires little labour to get on a truck from less dagging, drenching, dipping and now shearing. Industrial activities are characterised by the substitution of labour, either by replacing with machines or separating into easily repetitive tasks. The beauty of working with nature is that labour can be reduced through genetics. An unspoken challenge is creating an animal that anybody could start organic farming with.

As strong wool values have crashed, shedding sheep have become more popular. Through his UK work, Daniel recalls how the shedding fleece trait gained momentum. In 2012 a one-off UK Texel ram bought for £210,000 started peeling its fleece and overnight the stigma of shedding was gone, especially for strong wool.

> Genetically, they aim for calving ease, high growth and plenty of marbling to calve progeny in Australasia's top 1%. – Colin and Keryn Brown

Sheep shed at different times in response to day length. It's usually the same ones that shed earlier but this year (2023) his earliest shedding ewe shed later because of raising triplets, so stress delays shedding. Ewes shed earlier and faster as they age. Shedding can be linked to sex traits with ewe hoggets shedding more than rams while rams are more prone to footrot. Crossing Wiltshires and Dorpers produces offspring which sometimes don't shed much wool. Daniel is aiming for a shedding Texel so now invests in UK genetics.

As a leader, Daniel brought 20 breeders together for the Shedding Sheep Improvement Group, a group working together to accelerate genetic gain in shedding flocks. Linking sires and sharing rams allows benchmarking for linked flocks to rank against each other.

Daniel's ewes are Finn, Texels, Texel crosses, Wiltshires and Dorpers. Most have the GDF9 gene which is one-third of Finn fertility and less aggressive than the earlier Booroola gene. As ram breeders develop a consistent animal by crossing the most fecund sheep, Finn, with the fastest-growing, Texel, the national market is evolving from half-half Texel-Finn crosses producing too many lambs to threequarters-quarter. Fertility and production are often at opposite ends of a see-saw.

Other labour-saving traits include shorter tails as farmers look to drop tailing and crutching costs. Another is worm resilience.

With intestinal parasites, breeders notice worm-resistant sheep are often fussy eaters. In another major project, the Beef and Lamb NZ Genetic Low-Input

Central Progeny Test, one ram's daughters had low worm burdens but low growth on pasture. When scientists changed diets from pasture to lucerne nuts, to do feed efficiency studies, the ram's daughters were slow to adjust to the new diet compared to others but once adjusted they ate well and lifted growth rates to nearly the best in the group.

A breed example is how Finn liveweight gain is slower than Texel as they'll graze down to a residual in three-and-a-half days while Texels do it in two. This is why Texels out-grow Finns but are susceptible to worms, creating a production versus animal health see-saw.

> When farmers experiment, they change several things to accentuate results because it's outcomes that matter – Jeremy Casey and Kim

Breeds also have differences with trace element consumption. Texels originally come from copper-deficient islands whereas Finn derives from places with higher copper levels. Therefore, farmers must be careful if both are in one flock when drenching copper as what's enough for one might kill the other. As Finn has higher tolerance to copper, they do better with Texel.

Being an artisan, Daniel achieves scale without capital by securing peppercorn rentals. In focusing on genetics, he's looking to reduce labour and lift productivity. However, as researchers around the globe know, not everything can be lifted at once.

Eco-profit Observations

- New Zealand feral sheep breeds offer hardiness insights.
- Peppercorn leases can cheaply build livestock numbers but time and distance can compromise grazing management.
- To reduce labour, productive sheep require less dagging, drenching, dipping and now shearing. Developing a ram stud offering these traits lifts animal value.
- The stigma of shedding sheep disappeared after a UK Texel ram bought for £210,000 peeled its fleece.
- Day-length, sex, age and stress all affect shedding of wool.
- Fertility and production are often at opposite ends of a see-saw.
- Other desirable traits include short tails and worm resilience.
- Highly worm-resilient sheep only graze the top third of the plant, never lower.

51. Angela Payne - Profitable Carcass Products

Angela Payne has developed a unique business proving offal products are the most lucrative parts of carcasses. She set out wanting to manage farms. Now she owns several.

While born and raised on a farm, she could never land a farm job interview. Nobody (including Smedley Station training facility) would take on girls in the 1980s so Angela's farming career started as a vet nurse. She credits Viki Bryson as the first Hawke's Bay girl employed as a shepherd, then farm manager, in breaking the glass ceiling for women in ag.

As a vet nurse, her first interest was parasitology. That led to a Kelloggs leadership project challenging accepted protocols about drenching and parasite management. When her meticulous research delivered results favourable to product promotion, it was publicised. Anything not was silenced. That hypocrisy underlies her opinion many animal health products such as some newer vaccines and drenches are a rort.

> The facilitator was so impressed he asked other DairyNZ staff to visit. None did, so in frustration he quit and went back to being an accountant – Peter and Mandy Paterson.

Vet nursing opened a door. At home with young children, she was asked to go into abattoirs to look at extracting tissues because of her background knowledge of anatomy, glands, parts of organs and specific tissues. The idea was to work alongside butchers on the chain to extract the parts not being saved. Her business Agri-lab started there.

Despite being a busy mum, contracts for brain glands, dissecting bovine foetuses, harvesting tissues onsite and collecting products all fitted easily with raising a family. Being on the spot, supervisors would ask her to collect this or that. Then other companies began ringing up asking for products to collect. Initially Angela worked with 5-10 products, wanting 5-10kg of each, which then created opportunities for meat companies to ship off three container loads of something else. Being in the right place at the right time created more sales and referrals.

But working on highly structured worksites has issues. For example, collecting thyroid glands from severed heads. On the chain, if necks are cut in the wrong place, product disappears.

Then labour issues, for example collecting adrenal glands. They can be plucked from carcasses while being inspected. But when labour is unionised, each job is strictly described, meaning if a job falls between positions it requires a whole new person. If there is no space for that person on the chain, then that's another opportunity gone.

These problems led to developing her own facility. Angela built Agri-lab into an offal-processing and freeze-drying business. Buying and dissecting offal in her own facility enhanced harvesting efficiencies.

During dry-aging, enzymes break down connective tissues, tenderising the meat and intensifying flavour. Shrinkage is 8% – Jamie and Nicky Gaddum and Robert Hesselsteiner.

The business grew from market demand, not product push. Placentas were her first winner, exported to cosmetic companies as ingredients for anti-aging dietary supplements. They are fermented to breakdown proteins into simple amino acids in a digestion chamber. This makes products easier to grind into powders, otherwise stringy fibres from connective tissues would contaminate them.

Each country wants different placentas. Japan prefers horse and pig, US will take deer, sheep and horse, Australia and Canada want sheep, China and Hong Kong prefer not just deer, but wild deer. Luckily, not everyone wants the same thing.

Another interesting product is extracted bone marrow sent to Germany for drugs to alter DNA as part of RNA therapies. These products do exist; hence mRNA vaccines can change DNA. The full order might be only 5kg.

Overseas markets want products that are GE and BSE (mad cow disease) free. They know the value of natural products and that New Zealand is unique with its GE-free status, something now under threat.

Crucial for her business are freeze driers for blood and placenta products. Buying driers increased cash flow from processing more products, such as deer milk for Korea, blood for Australian black puddings or horticultural lines such as kiwifruit powder for digestion aids.

The business has unique qualities as it handles many condemned products and has a special licence for handling dairy products as they are usually kept separate from meat products. In looking to retire from the business, Angela believes any buyer would gain an instant return of 10% in the first year on current turnover, and significantly more soon after due to its huge potential for growth.

At her Waipukurau base, Angela initially invested in residential and commercial property. Her strategy was to buy and hang on because even the worse suburbs become flavour of the month. Some houses increased in value by 600% in 15 years. Recently, her focus shifted to rural property around Waipukurau but also further south with the purchase of a high country station at Hakataramea, South Canterbury, and a vineyard at Tarras, Central Otago.

For those looking for success, Angela's advice is:

For farm management, focus on every day being a growing day for either grass or stock. You need to prioritise actions when you've got so much going on. Sometimes only work with 85% of the job if you can get away with it, sometimes 95% or 100%. You need to balance getting things done against perfectionism.

What is a worst-case scenario? Once you've faced this it usually doesn't happen. But if it does, you've thought it through. There's power in fixing mistakes. Get gates moved or put in new ones, make life easier. If an action is going to avert a short-term crisis, get on to it.

And another thing. When it comes to handshakes, never crush a woman's hand. That deal and any other is over faster than you can ungrasp.

Eco-profit Observations

- When her first career path didn't materialise, an associated path led to greater rewards.
- Research into parasitology raised suspicions about commercial ethics.
- Visiting abattoirs created enterprise opportunities by just being there. It meant her business was driven by demand.
- Looking for business opportunities within highly structured processing facilities designed to maximise efficiency requires a guerrilla tactics mentality.
- Building her own facilities provided flexibility to optimise harvesting efficiencies.
- Placentas fermented into amino acids for dietary supplements was her first winning product. Every country wants different placentas.
- Blood products used in RNA research for compounds to alter DNA.

- Freeze driers opened new markets for freeze-dried products such as in horticulture.
- Investing profits into property allowed her to capture market appreciation values.
- Don't need to be a perfectionist if you can get away without being one.
- Plan for worst-case scenarios as at least you've thought it through.
- Never crush a woman's hand when shaking on a deal.

52. Market Cycles and Price Swings

The events in the life of a cycle shouldn't be viewed merely as each being followed by the next, but—much more importantly—as each causing the next. Howard Marks (2018). Mastering the Market Cycle: Getting the Odds on Your Side.

I've yet to meet a New Zealand farmer who truly understands livestock cycles. In his book Knowledge Rich Ranching, Allan Nation explains the North American cattle cycle lasts roughly 10 years. His point is that knowing its stages helps ranchers manage risk by forewarning of costs and prices. Most livestock producers haven't a clue and therefore are open to exploitation. His observations are a great lesson in supply and demand economics.

Is rainfall infiltration and its relationship to kgDM produced ever mentioned when talking about production efficiency? – Valuing Rainfall Efficiency.

The reason market cycles occur in mature industries, such as the North American beef market, is demand for increasing supply tanks, then cycles peak.

As markets decline, rancher deaths and retirements lower both supply and the breakeven price for those remaining, thereby increasing profits. This eventually attracts new players to the market while encouraging existing producers to expand. These people want heifers, which lowers beef supply further and attracts more investors. This market behaviour conspires to lift prices higher and attract more people into the cycle.

But when demand stagnates, over-supply eases prices triggering the slaughter of cull cows and increasing the beef supply while pushing prices lower. Market focus shifts from cows, and particularly heifers which are required for expansion, to buying steers and bulls. This allows ranchers to ride down the market using higher rates of gain to outcompete falling prices. Nation suggests putting heifers in with cows for slower growth while the best feed goes to steers for maximising growth rates and returns as the beef price declines.

Ideally, a simple strategy is to buy as many as animals at the cycle's floor and to sell them at the top. However, there are number of issues as the market price rises. When sale prices increase, that lifts replacement values, which dries up cash flow due to higher taxes and replacement costs. So as herd inventory estimates rise, farmers are strapped for cash to cover higher costs driven up by lifting livestock valuations.

Despite a looming cash shortage, it's best to expand at the bottom of the cycle when calves are cheap and ranchers can take maximum advantage of re-investing profits as markets climb. Once the market peaks, then you cash up your cows, keep young stock as replacements and switch from trading heifers to steers to ride the market down. Allan Nation believed this was one way young farmers could raise enough capital to buy a farm by capturing livestock value before they depreciate.

Cashing in only happens when breeding stock are priced well above the floor meat price, making them riskier prospects than slaughter animals. In New Zealand we are seeing this with Wiltshires, where both ewes and ram lambs are selling for much more than the meat price. It also happens with any new breeds sold without much history before settling on their real mean price. In the 1980s there were 14 breeds of sheep in New Zealand, today the number of pure breeds is double, plus all the composite breeds. How many on the market are being oversold?

A trigger point to increase livestock numbers is a rising value of gain which reflects a thinner margin, whereas a widening margin or declining value of gain is a sign to stop increasing numbers and consider switching to heifers.

So while there is a cycle, what is happening with price? Nation points out the only real price is what a beast gets at slaughter. Everything else is due to human greed and fear, all emotion. Look at livestock studs. Stud farmers might insure their prize bull for $100,000 but what does it bring on the hooks? It'll be the same price as everyone else's.

Market price only reflects slaughter prices on the day. While steers generally keep their value as they get closer to an abattoir's door, heifers are often valued beyond their meat price because of their breeding stock value. The same occurs with yearlings versus calves, with yearlings becoming preferred over time because of less sickness, death and shorter finishing time making for better cash flow for feedlots. All this reduces risk.

Furthermore, knowing the average market price is important. This is the average producer's breakeven price, the price markets always return to as momentum stalls one way or the other. Farmers need to produce livestock well below the average or mean price because if they can't, profits will never be guaranteed.

Owner operators have more flexibility to substitute profit for experience but require wider margins to value resilience over revenue – Challenging Farming Culture.

Another important marketing term to understand is the "floor" price at auction is the ground meat price. Nation explains everything above that is "air". When meat markets crash, it literally means the balloon's popped and "air" escapes as breeding animal values plumet to the market meat price. But if markets fall through the "floor," it means farmers don't even get the meat value of an animal. And then they pay to bring it home.

So how do we minimise risk in New Zealand? The simplest way is to run sheep and beef cattle as they have almost opposing market cycles. Another benefit, according to Massey University Professor Nicola Shadbolt, is these cycles are long, 10 years. She says beef cows are most economic when used to create high-quality grass for another livestock class. Dairy on the other hand, has only a four-year cycle, hence higher risk, often compounded by the financial structures dairy farmers employ and a lack of other income-generating enterprises.

So what's the earliest indicator of market direction? As Allan Nation states, the steer price. When the rising steer price breaks, then heifer prices break too, though that'll be a year or two away.

Eco-profit Observations

- Mature industries have price cycles because demand peaks and prices fall.
- As markets contract, lower supply and rising prices bring more entrants. As demand increases the price for heifers rises restricting supply further. This lifts prices and attracts more entrants.
- Eventually, over-supply eases prices, triggering the slaughter of cull cows and increasing the beef supply while pushing prices lower.
- Put heifers in with cows for slower growth and funnel best feed to maximise steer growth rates.
- If buying many animals at the cycle's floor and selling at the top, increasing taxation and replacement costs will squeeze cash reserves.
- Cashing up a beef herd at the top of the cycle is one way for young farmers to raise capital to get into land ownership.
- A trigger point to increase numbers is a rising value of gain which reflects a thinner margin.

- The only real price is what a beast gets at slaughter. Everything else is due to human greed and fear, all emotion.
- Heifers are often valued beyond their meat price because of their breeding stock value.
- Farmers need to produce livestock well below the average or mean price because if they can't, profits will never be guaranteed.
- If markets fall through the "floor," it means farmers don't even get the meat value of an animal.
- New Zealand farmers can minimise risk by running sheep and beef cattle as they have almost opposing market cycles.
- Earliest indication of market direction is the steer price.

53. Bryce and Aneta Lupton - Secrets to Trading Success

Bryce Lupton's advice about trading livestock is simple: "Stick to basic policies, don't chase markets, it just works. It's all about making money with good pastures, no need for fancy pasture species or mixes. Work with what you have. Keep it simple. No silage, hay, supplements, pasture renovation or fodder crops. Only graze animals when you have grass."

Up to 2000, Bryce was running a traditional Coopworth flock of 3,000 ewes on his family's 385ha block at Marohemo, Northland. Today, with leasing a further 135ha, he winters around 3,500-4,500 lambs and anywhere up to 950 yearling cattle, mostly bulls. He avoids dairy heifer grazing as that means making supplements.

The trigger to change came from being the local Meat and Wool NZ monitor farm and in discussions with its associated farmer network. He started looking at farm management and production data more closely and decided to test the limits. He tried 100 cows and 1,500 sheep. Then he crossed Charolais over Hereford-cross Friesian cows but struggled to make a dollar.

When he realised growing animals generate most income on livestock farms, conversations got him looking at the "asset," how to cheaply grow lots of grass. While the asset is total pasture biomass, it's how much grass each livestock class consumes that drives change, particularly comparing productive versus maintenance classes. When ewes eat the bulk of grass, production and income from lambs remains limited.

Furthermore, a Farm Environment Plan got him looking at soils to protect them from erosion. The farm's soils cover limestone, some only 15cm deep, therefore light feet are ideal. This led to a change in stock policy to lighter animals to avoid winter pugging and stimulate earlier grass growth. This is the advantage of sheep, plus they thicken swards too.

> Dropping superphosphate and potassium fertilisers resulted in a surprising absence of flystrike despite a high fly season – Andrew Tabener.

Other insights also emerged. During monitor farm days he observed new pastures on other farms only lasted three years, the result of set-stocking and overgrazing. Now he'll happily go with natural seeding. If pastures don't seed, paddocks end up full of weeds and need expensive and risky re-grassing.

Fertiliser became simpler too, just super and lime. He tried DAP to push lambs along in autumn, but found he needed grass to start with. Now it is too expensive with maintenance rates dropping from 25kg/ha to 20kg phosphate with sulphur and molybdenum for clover if needed.

But with livestock it was another simple insight that energised change. Winter growth rate and prices are high while summer growth rate and prices are low, a secret to success. It's Bryce's unfair advantage in the commodity game.

In switching to finishing winter lambs, scale creates another challenge. Small mobs of local lambs were too variable so now he sources them from the "Forgotten Highway" (Stratford-Taumarunui). Lambs need to cope with an 8-10-hour truck ride and if they don't, Bryce happily sends them back. As he mentions, "It becomes expensive for suppliers when sending two or three back and makes livestock reps take notice." All are quarantine drenched on arrival.

> Longer rounds result in cows consuming a higher fibre diet which reduces many animal health problems – Karl Steiner.

Bryce buys lambs from March through to May depending on the season, then drafts off every week from the beginning of July. Ideal lambs are 32-33kg liveweight to sell at 48kg to get 20kg carcase weight. Simple. If it's a good season, he'll push from 20 to 21kg cwt as it is very easy to get that gain. Lambs are sold within 95% of buyer's specs.

As lambs leave, in come autumn-born bulls. Bryce observes the roughest calves grow fastest. They've been weaned off meal, which hastens rumen development. Problems come from calves straight off meal struggling to adapt to pasture, so he prefers buying calves fed less than 0.5kg meal/day. It helps if they're 110kg lwt as their rumens are better developed than at 100kg. All are drenched with Turbo and get a pour-on for ticks on arrival.

Switching species reduces parasites. Lambs clean up pastures for calves and cattle do the same for autumn lambs. In a good season, he'll take top bulls through to 530kg lwt, producing 270kg cwt. Keeping autumn-born calves through two springs instead of two winters, means heavier animals when soils are drier. In mid/late spring they'll be putting on 1.5-2.0kg/day.

However, this year (2023) he started with the best lambs and bulls ever but they struggled to gain weight. A cloudy wet winter and spring reduced photosynthesis so pastures contained little energy. He sold lambs early and short at 19kg cwt. Bulls struggled to gain weight too. Autumn-born bulls are grown at home while spring-born stay at the lease block for simplicity.

For bulls, Friesians are easy to trade on the phone with data and videos. Bryce has many repeat customers looking for stores, especially from Waikato. He has a simple protocol for weighing stock - he stands them in the yard for three or four hours in the morning before weighing them. Customers note weights are true. He sells store bulls from December-on, depending on the market.

Over 35 years Bryce says the trick to trading is selling when animals reach budget value. He says it is best to sell then rather than wait as markets can drop 15% fast. And it is best to sell when everyone else has grass and be in the market first. To get there, it's best to destock early to grow grass for when others have little. Simple. He doesn't compete with kumara growers as after they've sown out paddocks they'll pay anything. Nor does he want to compete with suppliers in Waikato or Hawke's Bay.

Bryce believes his greatest achievement is farming the same land his ancestors have done since 1896. When you've discovered your unfair advantage, the path to success is clear. The trouble is keeping it simple. Flexibility is key.

Eco-profit Observations

- Growing animals generate more income than mature animals.
- Having productive classes consuming cheap grass became his focus.
- Lighter animals are better for soils. Sheep thicken cattle swards.
- Observed new pastures lasted three years due to poor grazing practices.
- Allows pastures to naturally seed so paddocks don't germinate weeds and need expensive and risky re-grassing.
- Fertiliser rates minimal to keep costs low.
- Focus on winter grazing as growth rates and prices are higher.
- Synchronises buying and selling of livestock classes to reduce parasite problems.
- Simple protocol for selling stock; stand them in the yard for three or four hours in the morning and weigh after that.
- Sell when livestock reach budgeted targets. Markets drop 15% fast.

54. Robin and Lois Greer – Pioneering Organic Dairy Products

"Nothing is more satisfying than growing grass and turning it into a product people enjoy," says entrepreneur Robin Greer of Retro Organics. "Most New Zealand farmers never experience that. They never get the feedback they deserve."

Robin was originally an apprentice butcher in South Auckland. That experience opened his eyes to retail's potential for making a dollar, because once in dairying his aim was to sell dairy products.

That didn't happen overnight. A hydroponic operation allowed him to tinker with growing lettuces, herbs and strawberries in Southland's cooler climate while dairying. Later that experience proved valuable when selling dairy products to supermarkets.

He and Lois sharemilked in Waikato 12 years before joining Southland's dairy boom when buying an 80ha property near Gore in 1993. Both Robin and Lois lost their fathers to cancer while teenagers so removing the nastiest agrichemicals drove their farm management. They never had any intention of becoming organic because those farmers were weird.

However, upon building a factory in 2009, they realised ROI was higher switching to organic as their products would stand out in the market. They certified through Biogro to USDA standards, meaning no antibiotics. By now they had bought another 220ha property so they converted that, creating an outlet for organic cows treated for mastitis to align with the smaller, original farm.

> Organics fitted with farming and family values and they knew it was a better way both financially and for soils and ecosystems – Blair and Penelope Drysdale.

There are many growing pains with businesses. One is employees as nobody can do everything themselves. With Robin focusing full-time on processing and marketing, both farms were run by staff. While teams are passionate and great, few individuals make sacrifices for organic to truly reach its potential. For example, when personnel passionate for homeopathic remedies left, more cows drifted across from organic to conventional. They worked with that, but there's always the question of what could be possible.

Another challenge for new businesses is bureaucracy. Standards are often set by industry giants. For example, Fonterra gifted its Risk Management Plan manual to MPI. Those protocols don't work for small operators, often costing them out of existence. The Greers spent $50,000 on theirs. Strict adherence to one-size-fits-all procedures by MPI staff spawns many reasons for blowback against regulations.

An example is risk management plans. MPI demanded a second person signing off activities in a one-person factory. After pointing out the contradiction, Robin finally declared he'd carry on regardless and only then did MPI allow production to continue. The Greers found MPI staff highly trained but lacked commonsense and humanity. Robin says they have power and know it, and seldom have empathy for genuine businesses. Their focus is not on those they serve.

From 2010, Robin and Lois produced organic milk, Greek yogurt and up to 14 cheeses. When he was building the factory, Robin would spend one day a week in Lois' kitchen developing cheese and yogurt recipes. Clients included up to 300 retailers with cafes, restaurants, delicatessens and speciality and organic stores. They tapped into a niche market segment favouring traceability, the millennial foodie generation.

When starting up, demand and supply vulnerabilities are many. While product demand grew, Christchurch earthquakes knocked out their largest retailers, Liberty Market and Foodstuffs. That created losses. Foodstuffs further complicated demand consistency by moving to central distribution centres resulting in lumpy orders - half a pallet this week, five the next, one three weeks later, so they lost touch with weekly throughput on supermarket shelves.

Even getting shelf space was a battle. Their marketing struggled against Fonterra (Fresh n' Fruity) and Goodman Fielder (Meadow Fresh) until they received a lesson in having a point of difference. The idea of lactose-free dairy inspired a trip to Australia in 2012, where over a few days Robin learned to make yoghurt. They hired professional marketers who organised a TV news crew to pick up the story and demand trebled overnight as nobody else did lactose-free yogurt.

But this success bought problems. The business was consuming more than 20% of its milk supply to Fonterra. While local Southland staff weren't concerned, as it didn't affect them, Auckland's head office demanded it stop.

Luckily Open Country accommodated them right away but it revealed Fonterra's draconian attitude to expanding start-ups.

The factory grew to 23 staff and could process all milk from both farms, 2,000 litres/hour if demand increased. They delivered milk and yoghurt all year and cheese only over spring and summer. But expansion proved costly.

Upon picking up a contract for liquid milk in 2016, their major customer went broke and closed owing over $250,000 that was never paid. At the same time Robin ended up immobilised in hospital. Recovery took several months.

Those events made the Greers challenge what they wanted from life and to create more balance. They decided to sell or find an investor to ease financial stress. Despite having a ready market with products people wanted, neither of these options materialised so Retro Organics closed in 2017 and debt transferred to the farms. Today everybody is looking for such a business.

They picked up Open Country's organic contract but sold the smaller farm to pay down lingering factory debt. The larger farm of 200ha still runs 470 cows today with EU organic certification. By using chicken litter, pasture production is reaching nearly 14,000kgDM/ha, as much as they were doing conventionally. However, transporting it from Canterbury is making them look at their figures.

> The trend in Ireland and Europe is to restrict diets to 17% protein, with Irish research backing a drop to 14% and increasing milk production – Jeremy

Starting any business requires passion, tenacity and luck to overcome unplanned crises along the way. But knowing when to stop is also important. Ask Robin if the journey was worth it, and he beams "absolutely".

Eco-profit Observations

- Most New Zealand farmers never experience feedback they deserve.
- The experience of growing lettuces, herbs and strawberries proved valuable later when selling dairy products to supermarkets.
- Spread risk by converting the larger property while keeping the smaller property conventional, therefore had an outlet to send organic cows treated for mastitis.
- While teams are passionate and great, few individuals make sacrifices for organic to truly reach its potential.

- Strict adherence to one-size-fits-all procedures by MPI staff spawns many reasons for blowback against regulations.
- They tapped into a niche market segment favouring traceability, the millennial foodie generation.
- Central distribution centres create lumpy orders so lose touch with supermarket demand.
- Struggled for shelf space until a lesson in having a point of difference.
- Events made the Greers challenge what they wanted from life and to create more balance.
- Ended up selling the smaller farm to recover from their largest customer going bust.

55. Neil, Glenda and Teagan Gray – Reducing Raw Milk Risk

Ask Neil and Glenda Gray of Buttercup Dairies what triggered their interest in raw milk sales and they'll say the public campaign on dirty dairying. The Gray family has dairy-farmed at Orongo near Thames for 114 years. Neil read a magazine article on raw milk in 2017 and saw a chance to promote positive dairying while enlarging the 500-cow family business for daughter Teagan to take over. Their story is about minimising business risk.

Neil, like many of his generation, grew up drinking raw milk so he believed he had a customer base looking for a point of difference. However, showcasing dairy by selling raw milk is a contradiction. After all, the first question a doctor or nurse asks someone with an upset tummy is: Have you drunk raw milk?

Paranoia about raw milk traces back to North America's Swill Milk. During the 1800s cows standing in their own dung were fed distillers' mash, milked in filthy conditions and the milk coloured with plaster of paris and bolstered with water, flour, rotten eggs and molasses. It's estimated up to 80% of milk sales in large cities was Swill Milk, causing up to 8,000 infant deaths annually.

Consequently, modern milk is highly scrutinised and adulterated. If a single Waikato herd got bird flu, the Gray's raw milk operation would be shut down within minutes of notification. Dealing with extreme business risk necessitates simplicity.

Without any knowledge of market size, they kept costs low. A disused roadside milking shed, with its own driveway and parking, was an ideal site.

Supplying raw milk has been regulated since 2018 and requires meeting Fonterra standards. Renovating the 12-a-side herringbone shed involved building a new roof and lifting it 400mm, plus repurposing milking parlour equipment to save costs. Only the milk cooler was new. Milk must be cooled below 6C within two hours and delivered at 4C. The vat pump has to be surgical grade for body fluids. Two pods hold 300 litres each and are pushed 20 feet from the milk room to vending machines. Therefore, it's a low carbon operation as no transportation or processing is needed.

Customers see the milking parlour while filling bottles so image is everything when emphasising a positive story. With only 20 cows, they can run all cows into the parlour, leaving little dung in the yard and the tiny effluent system is easily emptied every five days. All cows graze nearby paddocks using Halter collars. The Grays like their small herd and enjoy cow to cow and cow to human interactions. As every cow has a personality and they talk to them often, so milking is very relaxed.

> All cows milk for 305 days, then get 60 days off. His highest-producing cow recently squeezed out 1062kgMS and his highest heifer 800kg – Cliff Shearer.

For biosecurity purposes they operate a closed herd. The cows for the raw milk herd are chosen in spring from their main herd. All are 4-5 years old, DNA tested to ensure they are A2A2 and have low somatic cell counts. Drawing cows from the main herd offers flexibility as they can add cows, or send them back, as demand changes. This is a key risk strategy if they are forced to shut by bird flu, for example.

As Glenda states, there's only two perfect foods, raw milk and raw eggs, but they're also perfect for bugs. They send samples every 10 days to Eurofins. After two years of clean records, they're now audited only once a month for pathogens so compliance costs have fallen by two-thirds. Clearly, a positive attitude toward compliance is necessary. Auditors aren't police. Neil's view is that compliance rewards best-practice and improves the quality of farming.

> Testing milk for Johnes during herd tests, vets found blood tests on high positive cows correlate with a milk test – Louise

The only time contamination can occur is when putting cups on. They spend more time cleaning cows than milking, spraying iodine, wiping it off after two minutes, then using teat-spray.

If an e-coli grade, or high SCC occurs, they give the whole herd a rapid mastitis test. E-coli mastitis can clear itself up very quickly, so if a first test indicates it, a test at five days will show nothing. Their SCC limit is 160,000.

Selling liquid milk only on-farm removes many compliance regulations and costs. For example, delivering bottles involves vehicle registrations and cleaning equipment audits. This means their customers are responsible for their own hygiene. As Glenda observes, security cameras show mostly men sniffing their bottles before shaking them out. It's interesting that customers prefer glass bottles to plastic.

There's no coffee cart onsite as regulations state each customer must sign a waiver and buy their own bottle of milk for their coffee. And they can't give schoolkids a free sample. They have one customer that has a bottle of their own milk with a local barista. The Grays appreciate many customers have compromised immunity and a single glass can rearrange one's gut microbiome. As Neil jokes, "Why bother with probiotics? Just drink raw milk."

Their whole journey has been an exercise in public perception. Organic dairyman Ray Riding told them customers only know three things about dairying - dairy farmers are horrible, brown cows are better than black and white and dairy kills off rainforests by importing PKE. The Grays don't feed their cows PKE but a conversation recorded in their visitors' book where customers mistook a pile of wood chips for PKE reminds them how easily patrons jump to conclusions.

Semi-retirement sees them working 365 days, but when customers say their kids prefer raw milk from Buttercup Dairies to Coca-Cola, who'd want to give up?

Eco-profit Observations

- Interest in raw milk sales triggered by public campaign on dirty dairying.
- Showcasing dairy by selling raw milk is a contradiction because the first question medical professionals ask about tummy upsets is whether raw milk has been consumed.
- If a single Waikato herd got bird flu, they'd be shut down within minutes of notification.
- Renovating a disused roadside cowshed with a driveway and space for parking kept costs low when market demand was unknown.
- Vending machines next to milking shed created a low carbon footprint.
- Image is everything - small herd, small facilities, graze nearby pastures, use electronic collars.
- Small herd reduces risk if a forced shutdown as it can be incorporated back into the main herd.
- After two years of clean records, they're now audited only once a month for pathogens, so compliance costs have fallen by two-thirds.

- Selling liquid milk only on-farm removes many compliance regulations and costs.
- Grays appreciate many customers have compromised immunity and that a single glass can rearrange one's gut microbiome.
- Don't feed their cows PKE but customers have confused a pile of woodchips for PKE.
- Customers say their kids prefer raw milk from Buttercup Dairies to Coca-Cola.

56. How Livestock Save the Planet

The controversy over livestock is tragic. We blame them, but without them we cannot reverse global desertification, we cannot regenerate the world's arable soils, and we cannot regenerate the vast areas in the process of desertifying. That simply cannot be done with anything available to science except livestock and holistic planned grazing, or something better when we can develop it.
Allan Savory. (2017). Solving "Wicked Problems." Acres USA. Vol 47, 56-62.

Farmers never hear how livestock can save the planet. No livestock agency or rural representatives in New Zealand provide examples of livestock enhancing soil properties or biodiversity. The complete absence of anything positive reflects a vacuum of understanding regarding ecology and New Zealand agriculture. The lack of paddock observation skills linking ecology to agriculture drives this vacuum.

When he wanted to measure ecological improvements on his land, especially from livestock grazing, he enlisted the community – Finding Beneficial Relationships.

It's never easy to look at ourselves so let's go somewhere else, Nevada, USA. More than 20 years ago, I began reading of a couple doing remarkable things with livestock. Mining companies hired them after spending millions of dollars on technologies to colonise tailing piles and bring biodiversity back from a moonscape. When technologies failed, Tony and Jerrie Tipton were the last call and yet every time they completed a project their work astounded everyone.

Unlike traditional approaches to mine reclamation involving huge machines, agrichemicals, synthetic fertilisers, irrigation, plasticised mulch and fancy seed mixes, this couple would survey the site, assess how much hay, straw, temporary fence and stock water they needed. Then pitch to the mining company the costs of this equipment, plus cows and their management fee.

The Tiptons would camp on site and over several weeks feed and water cattle. After consuming all the feed, cows were sold off, money returned to the mining company and then everyone would wait – about six months. In that time a thriving grassland emerged and would remain for years.

This is remarkable because in some instances these tailing piles had a pH of 2, containing lethal cyanide, salt crusts and often no topsoil. Furthermore, these environments are in locations with erratic rainfall. Hardly an inch falls while cattle are there.

And yet, despite the Tiptons being published in environmental and grazing magazines, scientific papers, presentations to federal and local politicians and bureaucrats, you will have never heard of them.

What surprises all those that write about them is how little interest the world shows in remarkable outcomes they generate. There's never been a documentary on their achievements, no interviews with Sir David Attenborough or other ecology celebrities. Greening the desert using livestock is invisible. You'd expect the United Nations, government agencies, environmental NGOs and the whole world would be banging down their door. None do.

The reason it works is very simple. In these brittle tending environments where soil surface moisture is largely absent, microbes deposited in dung and urine help repopulate baked and frozen soil surfaces to drive the decomposition processes once moisture arrives. Note, trees don't do this.

The role of herbivores in brittle tending environments is twofold: Substituting functions of soil microbes by partially breaking down vegetation in their rumens, the only consistent source of moisture to drive decomposition processes, and secondly, stimulating landscape processes by seeding microbes with their dung and urine.

The Tiptons did this by spreading native grass seeds about, then covering that with the hay and straw that the cows ate. Covering soil surfaces with dried forage assists with maintaining soil moisture. As moisture collects in rock crevices and cracks, trampling triggers germination of this seed source by ensuring surface contact. Germinating roots exude carbonic acids to solubilise rock for minerals, and a grassland emerges. It also provides cover for insects and invertebrates to move in. Five years later it's attracting wildlife of all kinds.

In a world where biodiversity loss is shamed and regeneration of biodiversity is celebrated, how come nobody hears about these successes? They're not hidden away, it's not rocket science and anybody can do it, yet it's never mentioned. Writers like Dan Dagget explain this is due to prevailing beliefs of environmental organisations that humans can only ever harm the planet.

Humans have no capacity to use ecological relationships, like those regarding livestock, to heal landscapes. The only tool for environmental health is rest by removing all disturbance.

> It's well established that sub-optimal grazing causes pasture burnout because carbohydrate reserves in plant roots and crowns never fully replenish from severe grazing – The Cost of Overgrazing.

This folly is easily seen on Google Earth. Look along almost any high-country boundary with DoC land. One side is bare soil and native annuals, the other is a thriving tussock grassland. That thriving tussock grassland absorbs more rainfall, sequesters carbon and has more biodiversity due to longer seasons. And it pays rates and taxes, the DoC side doesn't. The disturbance by livestock increases biodiversity in this climate as evidenced by many high-country sheep exclusion sites.

As always, none of this is new. Percy Smallfield in The Grasslands Revolution in New Zealand wrote how Ruakura Research Station employed similar techniques to get pasture established on pumice and dried-out peat soils because traditional approaches of cultivation and fertilising produced unsatisfactory results. They cut hay from the best pastures and fed it out heavily on poorer ones. This was 60 years before similar ideas were expressed by Dr Peter Sears to the New Zealand Ecological Society in 1962.

Allan Savory, the only wildlife ecologist stating over 60 years livestock can save the planet, shares another technique using similar principles, called patch burning. It's where a patch of grassland covering the same area as a mob would gather on is burnt. Livestock lick up the ash for minerals while dunging and urinating to inoculate the site. The burn boundary creates an edge effect where biodiversity naturally lifts between two ecosystems. Patch burns help break up the monotony of landscapes.

Why do farmers continue to endorse agricultural leadership incapable of explaining and demonstrating to the public how livestock save the planet? Livestock save the planet by helping humans reclaim barren sites through inoculating with rumen biology. Furthermore, they can reduce fuel loads in forestry, create seeding sites on landslips and clear weeds from riparian areas, public walkways and waste residential areas. These examples clearly demonstrate how livestock and their ecological relationships can repair landscapes.

Eco-profit Observations

- No New Zealand livestock agency or rural representatives provide examples of livestock enhancing soil properties or biodiversity.
- Tony and Jerrie Tipton, Nevada, assess mine tailing sites for how much hay, straw, temporary fence and stock water they require to stimulate biodiversity when standard technologies fail.
- After six months, a thriving grassland emerges and remains for years.
- What's surprising is how little interest the world shows in the remarkable outcomes the Tiptons generate.
- Microbes deposited in dung and urine help repopulate baked and frozen soil surfaces to drive decomposition processes once moisture arrives. Note, trees don't do this.
- Prevailing beliefs driving environmental organisations is that humans can only ever harm the planet and have no capacity to use ecological relationships with livestock to restore biodiversity.
- Disturbance by livestock increases biodiversity in this climate as evidenced by many high-country sheep exclusion sites.
- Ruakura Research Station employed similar techniques to get pasture established in early 1900s.
- Patch burning uses similar principles to regenerate biodiversity with livestock.

57. Max Nichol - Stock Density Success

Production economics and farmer experiences makes it clear, growing more grass is easier with a few large mobs. There are many reasons farmers do not do this in New Zealand but a number of farmers have made this work for decades. The other surprising thing is it's usually on hill country, not flat land.

Max Nichol inherited a sheep and beef farm of 1,600ha in the mid-1980s at Clarks Junction, Otago. The property has been in his family since settlement, some 130 years. It is typical hill country of the area with wide, flat, windswept ridges and steep sides.

His father had run multiple small mobs and set-stocked in all paddocks. That practice transferred all fertility to stock camps and high points, and produced lots of tag where animals hadn't grazed. Pastures were tufty as they never had stock pressure to even post-grazing residuals and favoured weeds liking high fertility and compaction like thistles.

By the mid-nineties the farm had grown to 4,000ha. To simplify handling, he combined ewes and cows, hoggets and heifer calves, and ram lambs and bulls in separate mobs. As stock numbers increased, he ended up running 12,000 ewes and 600 cows in one winter mob. In 2014, a video emerged online of drone footage showing the mob on a winter move. https://www.youtube.com/watch?v=ZYNpppuUFjo

When he first put the big mob together, his farm discussion group visited. Seeing them in a 11ha paddock they thought animals would die cause there wasn't enough feed. But they didn't realise Max was shifting them daily.

When shifting the mob, he says, livestock learn not to bunch and providing they are moved when they're not hungry they don't rush the gate. It normally takes 45-60 minutes to shift the mob through a 14ft gate and cattle are usually first through. Never use a heading dog or huntaway to hurry them along, he says. The only exception to that rule is when on larger, rougher blocks with deep gullies.

He would check the mob from a distance in the first four to six hours to see if they were sitting down and spread across the paddock. If they were bunched at the gate, it meant they were starving but he doesn't recall ever having to give them another break. He knew he'd left his shift too late when ewes would run under the bellies of cattle to get through the gate. Sheep would push cow dung long past the gateway.

Smothers are the big worry for farmers with large mobs. Max experienced three in his time, all after animals had gone through the gate, not during or before the bottleneck. The one smother he remembers was caused by a narrow culvert after the gate where animals had slowed and the rest had run over the top of them.

> Production changes were almost immediate with lambing lifting to 175% with 50% off mum to the works – Tim Hawke and Murray Weaver.

The value of the large mob became apparent in spring. Then he knew he was on track. Pasture growth was strong and even across paddocks. Campsites and tag disappeared. Hillsides grew grass, leading to better land utilisation. Evening up fertility across hill paddocks stopped animals parking and overgrazing where feed was best.

In snow he would put out half a dozen hay bales. The large mob had an interesting behaviour when it snowed; they would shelter in a dip on the hillside, not venture into the gullies where they might smother. To reduce any such risk, he drained swampy gullies.

When it comes to farm development, he believes best pasture growth occurs during the early years once out of crop from native country by capitalising on available fertility. However, another reason for declining pasture production, he believes, is decreasing rainfall totals over time.

The sheep breed he stuck with was Romdale. These were bred by crossing a Cheviot over Romneys to create a Perendale and then cross that with a Romney ram to create the Romdale. He then bred the ewe lambs with a Cheviot ram and found these animals handled the large mob well. Mixed-age ewes did fine but when the condition of two-tooths dropped away he knew they were not getting enough feed. There were times where they would struggle a bit when introduced to the big mob.

He stuck with black cattle, using a mixture of breeds that included Angus over Hereford, Charolais, Simmental, Limousin and Maine Anjou. Cattle were less troubled by the large mob than sheep.

With only one mob to shift, Max spent his time fencing. By the time he'd finished he had more than 200 paddocks with an average paddock size ranging between 10-20ha. He also installed a gravity-fed stock water system comprising of 40mm PVC pipe which had to be glued in six metre lengths. Water is what made the large mob possible because animals are unlikely to graze areas without water.

They use longer rounds to reduce fertiliser applications and weed problems – Mark, Judy and Keeley Pudney.

But the real success was how stock density improved soil fertility. From what they consume, livestock return through dung and urine at least 80% nitrogen, phosphate and sulphur, plus nearly 100% potassium. Using 2023 fertiliser prices, the large mob could squeeze out around $5,500 worth of nutrients per day; up to $2 million worth of fertiliser annually. That would've swamped Max's fertiliser bill many times over.

No fertiliser rep or executive ever mentions the financial value of nutrients passing through livestock. How many farmers fully capitalise the tool of stock density to target nutrient transfer to even up pasture production across their landscapes?

Eco-profit Observations

- Multiple set-stocked mobs transferred fertility to campsites and high points.
- Creating fewer, larger rotating mobs reduced fertility transfer.
- Creating one mob of 12,000 ewes and 600 cows significantly reduced winter labour.
- Shifting took 60 minutes through 14ft gates. No dogs.
- Smothers occurred after the bottleneck, not prior.
- Campsites and tag disappeared, spring pasture growth was even, better land utilisation and less overgrazing were signs it was working.
- During snow, the large mobs would shelter in dips on hillsides, not camp in gullies.
- Romdales suited the big mob. Two-tooths would struggle at times to gain weight.
- Proper stock water delivery essential. Glued six metre lengths of 40mmm PVC to create the ideal water system.
- The big mob recycles $5,500 worth of nutrients every day.

58. Guy, Ashley and Ian Coats – Simple Cattle System

"Legacy creates motivation, when that disappears motivation goes. I can't run the business by myself. It's the perfect property, but after 21 years we are moving on."

Guy and Ashley Coats, with son Ian, farm near Elsthorpe, coastal southern Hawke's Bay. Guy grew up in California and Hawaii and his dream of ranching led them from his family's Californian real estate business to a property in Montana. Guy's grandfather ranched in California, Missouri, Texas and Oklahoma and spending time with him in Missouri as a 12–13-year-old are some of Guy's fondest memories.

In their mid-30s, Guy and Ashley bought a ranch in Montana, a small freeholding with a large lease similar to South Island high country properties.

But they struggled. The growing season was only 90 days long and they spent long hours making lots of hay and feeding out. Marketing options were limited because, essentially, it was the same climate 1000 miles in every direction.

Then they saw stories about New Zealand in Stockman Grass Farmer and National Geographic. After a couple of scouting trips they decided to go for it.

> A range of benefits emerged from questioning a winter crop and overseeing a gradual reduction from 140ha to no cropping by 2020. This wasn't planned – Travis Leslie.

They figured South Island was too much like Montana - hard winters, irrigation, etc, so they focused on Hawke's Bay because of its beaches and lifestyle. They bought Waipuna Station at auction and then the next-door property six months later to create a 1,500ha farm of coastal hill country, mostly steep Class 6 and 7 land. As Guy recalls, "It was our ideal property. Coastal hill country, good fertility, only lime, no super and plenty of grass and clover."

It was a rough experience starting in 2004. Sold as a going concern, meant they jumped in with 80% sheep and 20% cattle. The sheep weren't doing well, with low lambing percentage and lambs that didn't fill out. And there was this local rush to get everything to store weights by December 1. They had 13,000 sheep including lambs and were going backwards. Then the drought of 2007 hit and lamb prices plummeted to a dollar a kilogram liveweight - $20/hd.

They began questioning their whole operation. Why do sheep? How could the operation be simpler?

Grazing R2 steers worked and their management was much easier. In fact, why keep stock all year?

They sold all sheep and started trading R2 cattle. By growing and deferring grass over summer, they overwintered 1,500-2,000 head for someone else to finish in spring. It was simple but challenging - no hay, no break-feeding. They moved mobs paddock by paddock. Allan Nation's book, Pasture Profits with Stocker Cattle, became their inspiration to create cheap profits. When grass and overhead costs are largely fixed, Nation states, highest profits come from buying animals with the highest value of gain.

Running a breeding operation blurs finances with one year running into the next. A seasonal operation makes identifying mistakes and success easier, because a farmer can pinpoint where profit problems lie. Allan Nation observes that 85% of costs arise in the first 30 days of owning an animal, and the longer the animal is owned, the more weight and profit is gained. This system worked for them but it relied on a couple of marketing changes.

Timing of buying and selling is key. By buying in March and April when markets are focusing on weaners allowed them to source R2s out of the spotlight. They came from everywhere, but mostly East Coast. Then they were sorted into four or five mobs by weight with a forward mob always ready to go.

Selling started in September and October with the aim of having all out by January, although usually a few scrubbers were left to sort out, maybe 10%. If it was dry at home they could normally count on North Island's west coast to have grass and keep sales buoyant.

They sold R2s at farm weight with buyers seeing their animals running over the scale. They found deducting 4% liveweight worked well with most animals around 550-600kg liveweight. Selling on-farm meant vendors paid for trucking, which eliminated another cost. Having 20 truck and trailer units backed up along narrow roads around Elsthorpe excited locals a bit.

Going into winter with a bank of feed creates flexibility because altering stocking rate is cheaper than buying hay, silage or winter crops on Class 6 hill country. It also allowed them to promote later flowering plants by lengthening summer recoveries. This pushed red clover everywhere. It spread on its own and grew no matter the climate. Its tap root broke up any pugging and kept its quality all the way into winter.

> He started planting trees after seven floods in two months in his first year□- Graham Smith.

Son Ian led the way with environmental initiatives, fencing off riparian areas and planting poplar and other trees for the ETS. Now he's pursuing opportunities in Canterbury with a smaller property.

What's left for Guy is overwhelming on 1,500ha. The expectations of policy-makers leave many farmers with heavy workloads and emotionally burnt-out.

Another social issue is being in tune with local culture and expectations. They were constantly reminded, directly and indirectly, that having no sheep and leaving high post-grazing residuals wasn't the sign of a good farmer. Hiring staff was challenging because they always wanted to do things the Kiwi way. Guy worried about what others thought, but a banker told him, 'Worry about what's happening inside your own fences'.

Allan Nation states, "People are happy to fail conventionally than be successful unconventionally." Average daily liveweight gain is set by post-grazing residuals; grazing to ground level sets a low rate of gain. With that in mind, Guy asks, "With today's prices why farm sheep? They leave no feed; it doesn't make sense. Cattle are a better option."

Eco-profit Observations

- Ranching in Montana tough due to climate and markets.
- Bought ideal property - coastal hill country, good fertility, only lime, no super and plenty of grass and clover.
- Questions traditional production systems, especially after drought and a market collapse.
- Why do sheep? Why keep stock all year?
- Winter-graze R2 heifers and steers, simple but challenging - no hay, no break-feeding.
- Highest profits come from buying animals with the highest value of gain.
- Seasonal operation makes identifying mistakes and success easier because he can pinpoint where profit problems lie.

- 85% of costs arise in the first 30 days of owning an animal.
- Bought when the market was focusing on weaners, allowing them to source R2s out of the spotlight.
- Sold R2s at farm weight with buyers seeing their animals running over the scale, and a 4% liveweight deduction kept everyone happy.
- Altering stocking rate is cheaper than buying hay, silage or winter crops on Class 6 hill country.
- Red clover tap root broke up any pugging and keeps quality all the way into winter.
- Policy makers leave many farmers with heavy workloads and emotionally burnt-out.
- Guy worried about what others thought, but a banker said, 'Worry what's happening inside your own fences'.
- Farmers are happy to fail conventionally than be successful unconventionally.

59. Peter and Christine Johnstone - Developing Dryland Dairy

Ever notice how farmers achieving more with less are found at the end of no-exit gravel roads? Peter and Christine Johnstone's farm has two.

Peter ran an intensive dairy operation in South Otago. Despite winning a national management award in his early twenties he was burnt out by 24. He married Christine and they left for Europe. Then they spent a decade in Africa doing missionary work.

They returned to take up 50/50 sharemilking in South Otago for several years. When a dryland dairy property came up for sale bordering Riversdale, Northern Southland, they sold their herd to buy stock and plant.

Peter describes it as niche - drier climate and soils on gravel. The truck and trailer carrying all their equipment turned around in the shed paddock mid-winter barely leaving a mark. Anywhere else in Southland it would have bogged up to its axles.

The 120ha property was in Southland's bottom 5% of production, averaging 79,000kgMS, with a low of 50,000 and a high of 91,000 driven by grain and urea. Pastures riddled with grass grub provided little winter grass, making for a tough first spring.

Combining their African experiences with a healthy scepticism of consultants, they were ready for new ideas. A low-producing farm offered an opportunity to accelerate capital values and equity. For Peter, a recovering workaholic, employing once-a-day (OAD) milking allowed him to spend time with his family (five children under 10) and community. The school is only 300 metres away.

Development focused on re-grassing 25% in the first year, 30% last year and 25% this season. Pasture production lifted from six tonne/ha to 8.2 tonne and an expected 9.5 tonne this season, all without urea. Current pasture recovery (February) is 58 days ensuring clover component remains high with a seed drop at 60 days.

To rejuvenate the poor cocksfoot-dominated pasture they bought and spun on untreated clover seed for cows to trample in and then stitched in red clover.

Their multi-species pasture mix includes cocksfoot, ryegrass, white, red and persian clover, chicory and plantain. They aim for 30-40% clover in the sward.

Any deferred spring grazing is not balaged, as Peter says, "What's the use of all those nutrients locked up in $60 of plastic. It simply delays their recycling." Instead, it's mown for cows to trample there and then. Buying feed lifts milk production while increasing nutrients and organic matter to very light soils.

As for fertiliser, paddock trials with urea, sulphur and sulphate of ammonia always proved sulphur performed best. With Olsen Ps between 16-48, any production increases using super don't come from phosphate. Doing whole-farm soil tests resulted in many paddocks requiring up to six tonne of lime to lift soil pH into the sixes.

Winter feeding is 7kg straw with 7kg fodder beet. Despite yields between 25-29 tonne/ha, Peter is uncommitted to continuing with it but enjoys extra production in April and May to push grass into winter. Each fodder beet paddock is sprayed with effluent twice through an umbilical irrigator, and conventionally cultivated but with no added fertiliser. Using effluent makes the costs of growing beet much more attractive.

To strengthen their herd, they mated everything to crossbreds in their first season as they suit OAD and Peter's farming style. The 230-cow herd currently produces 460kgMS/hd, requiring two kilograms of grain plus 0.5kg PKE. Lifting pasture production resulted in 85,000kgMS in the first year, then 104,000. They are now targeting 110,000 this season and aiming for 120,000, but will need more cows. They're aiming for 500kgMS/cow OAD with all indicators heading in the right direction.

Better genetics is driving production with only the herd's top 25% and all heifers mated to nominated bulls. The rest go to beef. As heifers average 425kgMS this season, any cow not reaching heifer average production is culled, usually 10% to 15% every year. Heifers are grazed off-farm and come home before calving.

Calving starts on July 28 with half delivered in eight days compared to their first season's 45% in six weeks. With in calf rates of 76% three-week and 89% six-week, it shows milking OAD gets more cows pregnant and increases days in milk. Now they have a 6% empty rate after 8.5 weeks' mating.

The first round is 57 days followed by two rounds at 22 days to get to December, then a 45-day recovery pushes out to 56 days up to the season's end. This holds covers in what can be a very dry place to farm in summer. Pasture roots go down 2.5 feet with soil porous all the way to an iron pan.

… don't feed sheep minerals because relying on supplements to get performance perpetuates a deficiency that increases costs – Forbes and Angus Cameron.

As for animal health, MUN floats between 16-24. Peter uses fodder beet lick blocks grazing crop and magnesium lick blocks grazing pasture. Also salt blocks are offered all season post-calving. While cocksfoot is high in magnesium, cows will lick through three blocks a day in some paddocks. He observes high cell count cows needing lots of salt.

A surprise plus is weeds (thistles, nightshade, barley grass) struggle under the new farming style. Interestingly, docks have become quite palatable with cows eating dock leaves when entering paddocks.

Not all problem plants have biological controls. Cattle trampling deals to giant buttercup by crushing its crown – Vince Payne.

Peter and Christine are people every community needs as a backbone. Committed to faith in God, family and community they're creating a business reflecting their values. With production lifting 30% in three seasons by sowing deeper rooting pastures, enhancing soil fertility, longer summer rounds and focusing on productive heifers, they demonstrate getting basics right is key to developing a successful, simple operation.

Eco-profit Observations

- Burnt out dairying in his 20s and turned to a decade-long missionary career in Africa.
- Niche property, free-draining soils, dry country, but low production.
- Once-a-day milking allows time for family and community.
- Pasture development lifted pasture production 50% in three years. Longer rounds ensure clover seed drop.
- Mows spring pasture surplus. All supplements bought in.
- Sulphur gives best returns as fertiliser. Applied up to six tonne of lime to get soil pH over six
- Only uses effluent on fodder beet crop.
- Milk production up 130% in three seasons. Genetics driving production drawing on heifers from top 25% of cows.

- Better cow condition contributes to better calving performance.
- Pasture roots go down 2.5 feet to iron pan.
- Animals use lots of magnesium supplements. High-cell-count cows need lots of salt.
- Many weeds are disappearing under the new system. Docks are the favourite pasture plant of cows.

60. Peter Coles – Grazing Kikuyu Key

Peter Coles's parents bought his farm, north of Dargaville, in the 1950s, straight out of development by Lands and Survey. There was little kikuyu grass evident at that time because of fencing and stock control. By the 1970s, it covered 10-15% of the home farm. Then they bought the 100% kikuyu neighbouring property. Peter took over in 1981.

With a five-year average EBIT of $53/su and farm working expenses of half that, it's allowing him to continue doing what makes most young farmers cringe - the same management as his father – Richard Thorp.

Now, it is clear to Peter that less intensively grazed farms, whether due to inadequate fencing or poor grazing management, succumb to kikuyu invasion fastest. So, grazing management is key for keeping kikuyu in check. Mulching is too expensive for Peter.

Over the last 10-15 years Peter has focused on other species to keep kikuyu at bay. Ryegrass doesn't work for multiple reasons, but mainly due to dry summers. If it gets dry after sowing, livestock can't graze it as sandy soils blow away. And pests such as black beetle and grass grub decimate ryegrass and clover pastures. When reseeding didn't work, kikuyu would also invade.

Peter has found ryegrass paddocks do not last five years due to the dry, whereas cocksfoot hangs on much better. Before three recent years of summer droughts, he was looking to get eight years from his cocksfoot paddocks.

Through working with consultant Gavin Ussher, he noticed cocksfoot and tall fescue recover faster from drought than kikuyu. Therefore, competition and shading keeps kikuyu out of mixed pasture. Cocksfoot has a flatter growth curve. It doesn't leap into production in spring like fescue. Cocksfoot also has more clover over time compared to fescue. Hard grazing encourages clover, but when mis-timed grass species start to disappear.

Two ways overgrazing occurs is having animals stay too long and graze out the crown and growth points, or return too soon which doesn't allow root reserves to recover from previous exposure.

Peter likes chicory as a perennial. Chicory was the first plant he found to control kikuyu and any plants that survived spraying with glyphosate. However, that also created problems. Stitching other species into chicory pastures during drought failed as chicory was too dominant. Furthermore, paradise ducks foul chicory areas which reduces cocksfoot and fescue growth. Chicory responds very quickly with rain, but summer drought on sandy soils severely reduces its ability to thrive.

Peter's preferred mix for a stable perennial pasture includes 20kg/ha fescue, 3kg/ha cocksfoot and 8kg/ha chicory. As these are cool season species, he sows mostly in autumn to avoid summer heat. He also favours grasses with endophyte and prefers Hummer and Quantica tall fescues. Safin cocksfoot is finer leafed with better palatability than older varieties.

Prior to recent dry summers, Peter averaged 13ha of new pasture a year. He has done 60ha in annuals when droughts are severe, often requiring mulching of large areas. Even when new mixes haven't performed well, they are still better than the previous pasture. Therefore, he looks for kikuyu paddocks to resow and achieve better returns. If kikuyu is frosted, he'll stitch in new species to ensure something is growing by spring.

Recent consistent summer droughts have changed Peter's pasture policy. He's resorted to planting annuals instead of permanent perennial pastures. But in a surprising twist, he fallowed several paddocks and realised these paddocks were the only ones he didn't resow.

The power of deferred grazing and exercising a little patience reduces costs. With nitrogen costs doubling in 12 months, allowing paddocks time to reseed themselves might create a better return on investment.

When managing grazing Peter tends to jump mobs around his farm, grazing paddocks that are ready rather than next door. This ensures he minimises overgrazing of his preferred pasture species and lengthening the time he gets from pasture mixes. He grazes cocksfoot and fescue mixes at 15 days as longer recoveries reduce quality and fescue becomes bitter. Some paddocks he'll graze at 10 days.

When grazing recently cultivated and sown paddocks, he's found poa annua takes over. When this happens, he mows pasture before cattle go in because they'll pull out cocksfoot and fescue. Mob size can range from 30-100 head. He uses bigger mobs to reduce labour shifting pasture breaks.

He tried organics for a while, with the assistance of Waikato consultant Bill Quinn. One thing Bill got Peter to do was analyse plantain composition and compare it against mixed pasture species. It showed how herbs draw up significantly higher mineral content from the same soil. Within two years of stitching in plantain, stock no longer needed drenching with copper, saving time and money.

> Animal health issues are minimal because there's no problems with copper deficiencies or facial eczema – Geoff and Barbara Croker.

He had been trading stock before converting to organics, so to close the loop he carried 300 beef cows plus replacements. He dropped organics after delays and other difficulties with bureaucracy and has since gone back to trading.

Peter's focus is to make money with livestock. When trading, he's looking for something a bit smaller to create a larger margin. While always looking out for a bargain, he's also found buying cheap calves hasn't always worked because if too skinny they take much longer to fill out.

Peter overwinters around 650 steers and heifers which creates a better lifestyle than bulls as its more pleasurable. He starts selling stock once grass production declines and tries to always have a couple of mobs ready to go, especially if it's looking to go dry.

The challenge with meat production is balancing pasture longevity with livestock production as one compromises the other. Peter's deferred pasture grazing demonstrates that. Grazing management has the biggest influence on pasture species longevity.

Eco-profit Observations

- Less intensively grazed farms, whether due to inadequate fencing or poor grazing management, succumb to kikuyu invasion fastest.
- Ryegrass unable to control kikuyu.
- Cocksfoot and tall fescue recover faster from drought than kikuyu. Competition and shading keeps kikuyu out of mixed pasture.
- Chicory controlled kikuyu and other plants that survived spraying with glyphosate.
- Unfortunately, paradise ducks foul chicory areas which reduces cocksfoot and fescue growth.
- Sows cocksfoot and tall fescue mostly in autumn to avoid summer heat.

- If kikuyu is frosted, he'll stitch in new species to ensure something is growing by spring.
- During recent consistent summer droughts, paddocks he fallowed didn't require resowing.
- Minimising overgrazing of preferred pasture species lengthens longevity from pasture mixes.
- Bigger mobs reduce labour shifting pasture breaks.
- Within two years of stitching in plantain, stock no longer needed copper drenches saving time and money.
- Smaller animals offer larger margins, however, buying cheap calves hasn't always worked because they take too long to fill out if too skinny.

61. Adrian and Pauline Ball - Demonstrating what's Possible

Adrian Ball admits he's a bit of an outlier, a small-scale owner operator with attention to detail. As a dairy and beef farmer and active at higher industry levels, his focus is on demonstrating how New Zealand dairy and meat products can maintain an international advantage.

He reckons the New Zealand dairy model hasn't done a good job in broadening its scope away from production. It hasn't taken account of whole-farm systems. His operation is in the top 5% of properties in Dairybase, with the dairy farm emitting 10 tonnes of CO_2 and the runoff five tonnes. With an average of seven tonnes, profit hasn't dropped.

After buying his brothers out in the 1990s, marrying Pauline and converting the property from beef to dairy, he intensified the operation. But he didn't account for unintended consequences. He bought in PKE before anyone else and ignored his increasingly compacted soils because he was making a lot of money.

In the early 2000s a pivotal moment arrived when he couldn't pass a medical for his dive ticket. The doctor told him he had to change his job, a wake-up call. He was only in his mid-30s, but running three farms, two of them dairy, was burning him out.

Within weeks he and Pauline had sold two farms and dropped a lease block, which cleared most of their debt. He began running and two months later completed a marathon. Then he did the Coast to Coast endurance race in a single day. Focusing on his health improved his life and he found that with less debt farming could be less intense. His ego could be satisfied by other achievements.

However, the farm still bugged him. He wanted to fashion a farm that went beyond dairy problems. His view was that the last 20% of milk was unprofitable and sucking that out created large environmental problems. It's responsible for the imported feed that, by-in-large, farmers have made no money from.

Looking into greenhouse gas issues got him connecting environment and production efficiency. "We're in a new era," he now says, "where every resource is questioned. That's why I've been winter milking for the past 20 years, to avoid wastage." With nitrogen issues looming, they sold up their shareholding in a Canterbury dairy operation. That money bought a drystock block over the road from their Tirau dairy farm.

Annual production has changed from 200,000kgMS to 130,000kgMS, plus 50 tonne of beef. Now all feed surpluses from the beef unit go to dairy and dairy calves are grown out on the beef farm. This reduces costs and emissions and steadies cashflow. As milk volumes drop over summer, the meat cheques increase.

They rear 60 Jersey bulls annually for R2 heifers and use natural mating after AI to be self-contained. Their bull lines only come from the top 20% BW cows – Michael McMageira and Smita Apte.

They stopped importing feed and introduced cheaper methods, like no-till, to grow surpluses. Some 15ha of maize is strip-tilled with 14 tonnes/ha of chicken manure, generating 25 tonnes. Removing atrazine lifted earthworms to 40 per spade. Counting and weighing cobs per plant is the easiest way to determine yield. Then he resows with a winter annual for an additional 6-7 tonnes to fully feed beef all winter. All maize and grass silage fully feeds dairy cows, which average 600kg lwt and they'll consume seven tonnes of DM/yr.

Pastures get three tonnes/ha of chicken manure, resulting in higher soil organic matter levels. A mate using chicken litter will blend lime with it to compost for two months, then spray bugs on it. He also adds salt in his fertiliser to fix problems like bloat.

When their oldest daughter was told by her commerce degree lecturer that pastoral farming would be extinct in 10 years, she challenged them about their sunset industry. That started an interest in the Ballance Farm Environment Awards, leading to a win in 2019.

Methane threats encouraged a closer look at production efficiencies. If methane links to kgDM eaten, then optimising gains in the first year and having animals gone is a must. Calves growth rates require 7-8kgDM for every kilogram of liveweight in their first year but that increases to 11-12 kgDM/kg lwt in the second year due to higher maintenance. With autumn-born beefies, they'll kill bulls at 13 months, everything else at 18 months, and lower the stocking rate from mid-spring to put weight on cattle over summer. To lower emissions further, they'll slaughter at 80% of potential.

Their unfair advantage is higher winter growth rates, so they kill livestock when nobody else does. The key is matching supply to demand than holding cattle back. Therefore, their focus is prioritising growing grass by avoiding overly long recoveries, no dead plants and opening up pasture. They tried straight lucerne but were told by a US guy it would tighten soil so they are now planting more red clover to cover a February feed pinch.

> There's a paradox stud breeders juggle: Every commercial operator wants cows that lay down fats but processors don't want back fat on carcasses as cutting it off adds costs and time – Tom Hargreaves.

To improve beef quality from dairy requires beef breeds. Charolais gain 100kg cwt more than Belgian Blue with 30% right Eating Quality (EQ) whereas Angus is 70% right EQ. Fat determines eating quality. Belgian Blue has no marbling and therefore doesn't collect premiums as it is too lean for higher cuts. Wagyu takes too long to finish, creating higher emissions and is therefore not efficient.

They put Charolais steers and heifers together for steers to go at 310kg. There's a $300 difference in income between heifers and steers, which begs asking whether heifers eat $300 of feed to reach their weights. After 30 years of observation, a positive harvest number creates profit, therefore every bite helps.

So, what's ahead? Adrian believes the key to profits and meeting emissions will see the genomic dairy herd grazing more aggressively within 3-4 years. Shorter dairy seasons won't help because there's less production covering overheads. With dairy costs climbing (2023), beef returns are gaining parity, meaning the possibility of more stranded stainless steel is real.

Eco-profit Observations

- Profit hasn't dropped with lower GHG emission numbers.
- Could see intensive practices were compacting soil but making money was his focus.
- Medical triggered lifestyle changes, trained ego to get satisfaction from things other than farming.
- Last 20% of milk is unprofitable and sucking that out creates large environmental problems.

- All feed surpluses from the beef unit go to dairy and dairy calves are grown out on the beef farm. As milk volumes drop over summer, meat cheques increase.
- They stopped importing feed and began cheaper methods like no-till to grow surpluses.
- Removing atrazine lifted earthworms to 40 per spade.
- Counting and weighing cobs per plant is the easiest way to determine maize yield.
- Pastures get three tonnes/ha chicken manure resulting in higher soil organic matter levels.
- Methane threats challenged efficiencies. If methane links to kgDM eaten, then optimising gains in first year and having animals gone is a must.
- To lower emissions further they'll slaughter at 80% of potential.
- Higher winter growth rates allows killing livestock when few else can.
- Belgian Blue and Waygu don't collect premiums or have higher GHG emissions.
- Intake efficiencies are important for profits.

62. BLAND PASTURES CREATE BLAND FOOD

Research from Switzerland demonstrates that weedy pastures produce up to 300 per cent more CLA (conjugated linoleic acid), which means that these plants have much higher amounts of parent Omega-3 and Omega-6, which of course are fatty acids, the richest source of forage energy.
Jerry Brunetti. (2012). Cows don't have Carburettors, Acres USA, vol 42 no 5

My mother told a story growing up at Garston, northern Southland. A local Irish family always left a chain width (7 metres) of weeds either side of fences in cultivated paddocks. The reason was to allow livestock to self-medicate and remain healthy. Like locals at the time, many readers would ridicule such a backward practice and yet modern science proves it sound.

Early last century, scientists knew herbs played an important role in livestock diets by observing frequency, intensity and timing of grazing. Welsh pasture trials during the 1920s and 30s revealed interesting grazing habits of sheep.

Frequency of grazing individual pasture species remained the same whether in temporary or permanent swards, whether better or weed grasses dominated swards and even in cereal stubbles whether an abundance of productive grasses and clovers or volunteer strike of weed grasses surfaced. Grazing frequency of herb species remained the same no matter which pastoral ecosystem sheep engaged with.

Sheep grazed herbs more often in winter than summer months. Even where there were ample young grass and clovers in reseeded cereal stubble, the amount of herbs grazed was higher than expected. The greatest frequency of herb grazing was on unseeded cereal stubbles.

All this suggests herbs were the go-to plants when grasses and clovers were below their best. Most palatable species were rich in calcium and tended to be much lower in dry matter than grasses and clovers, suggesting maturity had a role in determining desirability of grazed species as seasons changed.

Studies at Massey University's organic dairy in 2006 showed weeds had a greater amount of minerals than either ryegrass or white clover, so much so a bale of weeds had greater fertiliser value than one of 70% ryegrass and 20% white clover. Organic farmers often observe improving livestock health with increasing pasture diversity because each species has different nutrient composition. Lower animal health bills are an unfair advantage in the commodity game.

> Any farmer giving up spring cultivation finds they have time to push a wave of feed forward into summer – Leveraging Time Lifts Profits.

Primary industry now favours plantain and chicory but it ignores similar plants. For example, ever notice dairy cows grazing dandelion flowers first when turned into a new paddock? This often signals a protein/energy imbalance in the sward. Such behaviour alerts farmers to both current and future health problems. Dandelion is considered one of the most nutritious salad vegetables, it's a super food. This is due to its tap root pulling nutrients up from below where soil tests measure. Has ryegrass ever been dished up at a salad bar?

English dairy farmer and ley farming pioneer Newman Turner was a great advocate of mixed swards for livestock health and performance, so much so he suggested farmers forget soil tests and use weeds as their indicator of soil issues. Very little is published on the interpretation of weeds and their environment. Even the Illustrated Guide of New Zealand Weeds from New Zealand Plant Protection Society provides no clue as to weed establishment.

Turner experimented with ley mixes in the 1940s and 50s to find which species best suited his management and production. He observed farmers bought seed mixes based on price rather than what each plant contributed to soil and livestock health. That hasn't changed because industry experts consider these attributes irrelevant.

To find out which species cows favoured he planted 35 plots with singular cultivars of common grasses and legumes as well as herbs for cows to graze. Cows preferred sheep's parsley, plantain and chicory first, then preferred burnet, kidney vetch, sainfoin and alsike clover. In comparison, ryegrasses, other meadow grasses, clovers, and lucernes were only sampled, if barely touched. Deep-rooted perennial herbs ensured pasture mixes were productive and healthy as Turner's prize-winning livestock constantly demonstrated.

Fast-forward 70 years and retired emeritus professor Fred Provenza's work demonstrates the value of plant diversity for livestock diets. His work at Utah State University shows livestock get bored eating the same thing. It's why browse lines are so evident on shelterbelts because of cravings for different vegetation. Different plants provide different nutrients. If the livestock mineral bill is high, it signifies a lack of pasture diversity.

For livestock to self-medicate they need to fine-tune their diet. According to Provenza this requires three processes to occur in unison: Flavour-feedback associations when grazing, availability of phytochemically rich plants and opportunities in utero or early in life to eat nourishing combinations of plants. How can this happen with a standard ryegrass/white clover pasture?

Food flavours and nutrition are dominated by plant phytochemicals. Provenza explains a reduction in plant diversity is responsible for bland flavours and low nutritive values in meat and milk while processed foods are fortified, altered and enriched to mask poor nutrition. So now human feedback mechanisms associate synthetic flavours and buzz from high calorie foods with desirability rather than satiety and nourishment from proper foods.

Other actions included entering the black economy and selling over the fence. Farmers made sure they were feeding each other – Equity Crisis Lessons.

This is why highly dense nutrition farming has to be forefront in our current health crisis. But as the title of his book Nourishment: What Animals can Teach us about Rediscovering Our Nutritional Wisdom, implies, humans could be more observant in developing their nutrition skills.

In the book there is an extraordinary experiment conducted by Clara Davis with orphaned babies at Chicago University in the 1920s, all overseen by the US Health Department. For six years, 15 orphaned infants under strict conditions were offered 34 different wholesome and natural foods at each meal. Without adult guidance it was observed no two children selected the same foods at each meal, nor the same mix of foods from day to day (p.21).

These children cured themselves from rickets and other ailments. They demonstrated when we listen to our bodies, we have every chance to cure ourselves. It brings a new dimension to the saying ‘behaving like an animal’.

Eco-profit Observations

- Grazing frequency of herb species remained the same no matter which pastoral ecosystem sheep engaged with.
- Herbs were the go-to plants when grasses and clovers were below their best, especially winter.

- Organic farmers often observe improving livestock health with increasing pasture diversity because each species has different nutrient composition.
- Dandelion is considered one of the most nutritious salad vegetables, it's a super food.
- Englishman Newman Turner suggest farmers forget soil tests and use weeds as their indicator of soil issues.
- Farmers buy seed mixes based on price rather than what each plant contributes to soil and livestock health.
- Different plants provide different nutrients. If livestock mineral bill is high, it signifies a lack of pasture diversity.
- Reduction in plant diversity is responsible for bland flavours and low nutritive values in meat and milk while processed foods are fortified, altered and enriched to mask poor nutrition.
- Offering a range of free choice wholesome and natural foods in a research experiment revealed children cured themselves from rickets and other ailments, similar to observed behaviour in livestock.

Adaptation: Bland Pastures Create Bland Food. Stuff, April 30, 2018 and New Zealand Farmer, May 5, 2018, p7.
https://www.stuff.co.nz/business/farming/advice/103486338/john-king-bland-pastures-create-bland-food

63. Hamish Browne and Robyn Budd – Deliver Brand Integrity

The best way to stamp your values in business is to have a brand. The challenge then is to ensure your product's integrity rides the whole supply chain.

Hamish Browne and Robyn Budd run Black Pond Farm, a 106ha organic holding in north Waikato's Mangatangi rolling hills. Hamish is a mechanic and Robyn's career is in advertising. Both come from rural and farming backgrounds and previously had small blocks at Karaka, South Auckland, where Hamish serviced equipment for construction contractors. For security, they continued with the mechanic's business to help with cash flow, although it is financially separate from the farm.

They chose beef for a farm enterprise because they both love cattle. While initially unsure if certification was worth the expense, five years later they are organically certified to Biogro standards for selling meat. This also aligns with USDA standards.

Everything they do on-farm is designed to strengthen their brand's values of caring for land, livestock and customers. A third of the property is in QEII bush covenant, while their wetland has a space planted with poplars to feed livestock in drought.

Another factor is improving soil function. Their land only absorbs 20mm of rain and any excess runs off. When saturated, soils seep water all the time and are a constant challenge for cattle, especially in winter. Their next project is building a stand-off pad to feed supplements during extreme weather.

Increasing rainfall absorption and water-holding capacity is a priority. Dung beetles help this aim. Their fertiliser regime includes sulphur, lime, dolomite, biodynamic preparations and a silica Ocean Elements mix. Silica helps with pasture photosynthesis and addresses high soil aluminium, along with lime. Interestingly, even with plenty of farm shade their animals do not seek it. Hamish and Robyn put this down to not using nitrogen fertiliser.

To give their livestock a grazing smorgasbord, they've resown new pasture mixes (19 species) with chicory, plantain, red and white clover plus various grasses. Deferred grazing allows pastures to seed, eliminating resowing as a square metre can contain 60,000 seeds. Hamish experiments with pasture species and mixes. They're trying new things all the time to move forward. About 70% of efforts work out and 30% fail.

Their herd is 40 cows - all Friesians crossed with either Herefords or Red Devons with 40 R1s and 40 R2s. They manage two herds, with a butchering mob of about a dozen. Every month animals are put over scales to determine weights and to decide which are for the chopping block. Being organic, they use homeopathy to calm animals for transporting, which works perfectly. Stock are processed from 24 months after animals start marbling, averaging two beasts a month in winter and four a month over spring-summer (Oct-Mar).

> The journey to low-input systems isn't easy because everything's geared the other way – Shane Armstrong.

When they started, no organic meat commercial stream from processing to market existed. Finding a top-notch butcher was easier than a processor. Everyone knows processors bottleneck the meat supply chain. Setting up a micro-abattoir was not an option because creating another business involved high start-up costs and time meeting MPI audits. Recently, a German visitor explained farmyard slaughtering of livestock in her country was much less regulated than New Zealand but labelling requirements were much more rigorous.

To keep in line with their organic animal production values, they freeze everything. With meat products, preservatives like nitrates are largely to stop sausages and salami deteriorating. Freezing does the same thing, allowing an 18-month shelf life. As far as MPI is concerned, they are a frozen and fresh food business, they don't touch anything raw. Frozen meat also allows on-sellers flexibility as sell-by-dates are longer compared to fresh.

All products are delivered freezer to freezer with meat box weights capped at 20kg, the largest a courier can handle. They've signed up to a logistics company and reduced courier costs. Many courier operations prefer clients with more than $5,000 worth of business a month.

The real challenge in growing a small meat business is what to do with non-premium cuts. The high-end cuts are easy to shift, ask any restaurant, but what do you do with everything else?

Some clients buy quarter and half-carcasses but their most popular are unusual cuts - for example, customers can't get enough beef cheeks and tails. Mince is their top seller and high-end cuts go to a restaurant in Wellington. As well as salamis, they have two sausages, keto and regular. Keto is 96% meat whereas regular uses a little rice flour. Regular breakfast sausages using sheep/lamb casings are crunchier than those pig casings due to their thinness. Skins are another opportunity they are currently exploring, either as hides or leather.

> Postage problems led them to buying a van for delivery runs to Auckland. Karen makes trips once a week, meaning they can ensure banana freshness and quality to customers – Ross and Karen Potter.

Their products sell in Waipu, Pukekohe, Drury, Thames, Hamilton, Whakatane and Rotorua. In supplying organic and boutique grocery stores, the lack of freezer space can affect economy of scale. They don't go to farmers' markets as running two other businesses, contracting and mechanical, plus the farm, means time is short.

Robyn does the advertising and mainly targets Instagram. There's no short cuts, so it means a lot of cold-calling for leads and opportunities. Being on Country Calendar was very helpful for contacts. They've produced signs and flyers for shops selling their products and do radio advertising at certain times of the year. Word of mouth is still best.

Hamish and Robyn tenaciously demonstrate their brand's values by growing livestock without modern pharmaceuticals, working with nature and creating products that don't require preservatives. Integrity is strengthened by being small, nimble and energised in response to changing market conditions.

Eco-profit Observations

- On-farm activities strengthen their brand's values of caring for land, livestock and customers.
- Increasing rainfall absorption and water-holding capacity is a priority. Releasing dung beetles helps.
- Silica helps with pasture photosynthesis and addressing high soil aluminium, along with lime.
- Animals do not seek shade, a phenomenon they believe is due to not using nitrogen fertiliser.
- Deferred grazing allows pastures to seed, eliminating resowing.
- They use homeopathy to calm animals for transporting, which works perfectly.

- Freezing all small-goods keeps products in line with brand values. Products have an 18-month shelf life so providing flexibility for on-sellers.
- The most-popular cuts are the unusual. Mince is their best seller.
- There's no short-cuts for advertising, just a lot of cold-calling for leads and opportunities. Word of mouth is still best.

64. John Vosper and Liz MacKay - Word of Mouth

As John Vosper signed contracts for a pasteurising plant and vending machines to set up and sell Jersey Girl Organics milk, what did their milk company do? It came out with a separate milk price doubling the premium for organics. As John puts it, “We began swimming like hell.”

The Vospers have farmed on the outskirts of Matamata for over 100 years. It’s believed the first local Jersey cows arrived when John’s great-grandfather, Cleave Vosper, walked them from Taranaki in 1911.

The home farm, Cleavedale, is 80ha with half leased from an uncle, including the dairy shed. It’s milking 162 A2A2 cows twice daily with calving autumn and spring. It will increase to 190 cows in 2024/25 season. Last year they leased a 120ha organic dairy farm 18 kilometres away with a herd of 280, of which 90% are full A2A2 cows, all spring calving. Nearby they own a 40ha run-off, as well as lease another near Mamakau for young stock.

A Californian lecturer triggered John’s interest in organics. He’d observed soil degrading under urea and pastures always needing more nitrogen. He started by dropping superphosphate for RPR. Attending a field day at Gavin Fisher’s inspired John and wife Liz MacKay to start organic conversion in 2002. By 2010 they were tired of their milk company’s attitude to organic production so looked at setting up their own brand.

Dairy companies rarely allow producers to keep more than 20% of production for peddling their own enterprises. To circumvent the 20% rule, their organic discussion group suggested setting up a second company with one leasing cows to the other. This has worked to date but some lumpy supply volumes have raised a few questions.

> Yet the 400-cow dairy made more money per hectare and per cow. The 1500-cow venture went for three years and dropping back to 1000 cows reduced stress for both himself and the contractor. He doubts it ever paid for itself – Roger Slee.

Their products are well received having won numerous awards allowing product merit to drive sales growth. They have many points of difference; A2A2, organic, no permeate or additives, no antibiotics or agrichemicals and reusable glass bottles at farmers markets.

While non-homogenised, it took a backpacker's comment for them to realise it's smooth and creamy texture was another strong marketing advantage. Nutritional benefits include protein averaging 28-30% higher than that stated on standard bottles. It also has 20% more calcium and higher omega 3 fats.

The greatest issue when marketing your own milk is balancing supply. Looking back, vending machines were probably the wrong market as keeping up supply in Matamata and Tauranga was problematic. Some people would travel tens of kilometres only to find vending sites empty. Supermarkets are more profitable because volume is higher and demand less lumpy. Their products are available in the North Island through selected Progressive and Foodstuff outlets as well as speciality stores such as Moore Wilson and Commonsense.

They service farmers markets in Tauranga, Hamilton, Grey Lynn and Parnell which are great platforms to educate customers on farm practices and politics. Many will buy other well-known brands thinking they're small operations only for John to tell them they're owned by overseas corporations, whereas the public will bitch about Fonterra not appreciating its owned by New Zealand farmers. Unfortunately, Fonterra is now selling many of its iconic retail brands such as Anchor and Mainland making customer connections to dairy farmers even more distant.

A benefit in processing their own milk was employing family. Today two of their three children, Michael and Laura, help run the business; Laura with finance and logistics and Michael on the farm and manufacturing yoghurt. They also employ John's nephew and niece to help with farmers markets.

All short-term financial gains from girdling stripped longer-term profits, something not revealed to unsuspecting new market entrants – Brett Petersen.

The problem with small factories like their own are costs associated with starting and ending shifts, including time cleaning, testing and compliance costs because smaller volumes drive up labour costs per unit sold. Now they only make yoghurt at home. Yoghurt is a line with huge potential because of increasing public interest in fermented foods. It also has a longer shelf life and a higher margin than milk.

To process milk cheaper they started working with Green Valley. Furthermore, joining with third party distributer, Big Chill, streamlined delivery and lowered packaging costs. However, shifting small volumes remains expensive. Professional logistics comes with a price tag.

Despite their success, they've barely invested a cent in advertising, it's all word of mouth. Farmers markets are the backbone of their marketing strategy by encouraging customers to taste their product; no billboard provides that experience. From there small shops and restaurants began ordering. They've gelato and ice cream venders using Jersey Girl's story as part of their marketing creating a national presence. Having truly reliable products founders repeat business with customers.

Like many dairy operations on the peri-urban fringe, dealing with costs and social pressures has the Vosper family diversifying into higher value foods. They've set up a hydroponic tunnel house for strawberries and other fruits such as blackberries, raspberries and passion fruit with their son-in-law and process engineer, Craig Jager. While non-organic, the berry business is small enough to ease into this industry by initially targeting farmers markets alongside Jersey Girl products with year round supply.

By highlighting texture and health benefits, Jersey Girl Organics is a niche dairy brand focused on taste and values. By targeting farmers markets, they've found a loyal following willing to be educated and nourished with their products.

Eco-profit Observations

- Observed soil degrading under urea and pastures always needing more nitrogen. Dropped super for RPR.
- Set up two companies, one to lease cows from another to challenge supply agreement limitations.
- Marketing focuses on many points of difference; A2A2, organic, no permeate or additives, no antibiotics or agrichemicals and reusable glass bottles.
- Nutritional benefits include higher protein, calcium and omega fats.
- Distance to vending machines made supply challenging. Supermarkets have higher volume and fewer lumpy orders.
- Farmers markets great platforms to educate customers on farm practices and politics.
- Small factories have lower throughput which drives up labour and other costs per unit sold.
- Fermented foods have a longer shelf life and higher margin.

- Moving to professional processing and logistics streamlined delivery and lowered packaging costs, but shifting small volumes remains expensive.
- All marketing word of mouth. Having truly reliable products founders repeat business with customers.
- On the back of their marketing success, they've set up another enterprise selling tunnel house produce.

65. Russell and Charlotte Heald - Fibre, Butterfat and Profit

"The biggest change is mindset," says Charlotte Heald. "It's deconditioning from industry beliefs and getting away from prescriptions. You need to slow down to observe." Husband Russell laughs, "Plenty of backlash going down this track. Apparently, we'll be growing trees, smoking weed and milking in bare feet." It's clear DairyNZ groups were no support during this time of change as they focused on numbers and ignored family and environmental wellbeing.

Yet in 2012 when they bought their Norsewood dairy farm, nothing could be further from their minds. They started farming conventionally and then a Tow and Fert got Russell into making slurries. That stirred his curiosity to ask more about soils. They met Greg Barclay, who pushed them over the line. Asking how to do things moved them into multi-species pastures and grazing management. Now they question everything. They started converting to organics in 2018, certifying in 2021, as it fitted with family values and food integrity.

Their 170ha milking platform runs 330 cows milking once-a-day (OAD) generating 330-370kgMS/cow over a 300-day lactation. The predicted 2025/26 season organic milk price is $12.30kgMS and by keeping feeding simple, their farm working expenses average $4.50/kgMS including wages. Compare that to Fonterra's standard milk price of $10 with DairyNZ's national farm working expenses averaging $5.84.

Milking OAD reduces livestock stress giving the cows higher resilience. With the longest walk 2.2km, it's easier on cows and creates more flexibility getting other things done. For example, with cups-on at 8am, it allows herd manager Sahra to breakfast with her kids before school and work around family life. After 3.5 years, Sahra is adamant she wouldn't be dairying without it.

While recovering from two years of drought and a production drop of 15% they are confident it will pick up. Meanwhile, they simply lifted profit by investing in a business coach.

Organics triggered a change to polled Jerseys and A2A2 six years ago. Jersey cows are known for their butterfat and the Heald's' herd delivers 7%, compared to Friesians at 3.5%. This is significant.

Twenty years ago, DairyNZ told farmers to only breed for protein, that fat was irrelevant. Increasingly, fat is becoming the target component of milk, as borne out by rising prices, while protein falls. When Fonterra announced its 2025/26 prices for milk fat, Jersey AI straws sold out overnight.

It's well known in nutrition circles, carbohydrates and fibre drive up rumen acetic acid production and that lifts butterfat, whereas the industry standard of excess dietary protein creates bloat, lameness and other animal health issues. The Healds make balage at 45-65 days, not 25-45 days, to lift carbohydrates and fibre. They buffer spring and summer pasture with 0.5kg/hd/d of hay to add fibre. They've found grazing diverse species also balances out excess protein.

He argues that by grazing "silage grass" he can milk an additional cow and a half every day for no cost – Henry Hendricks.

But sowing multi-species pastures delivers multiple benefits as they can either graze, cut or defer multi-species for later. By delaying grazing of some multi-species paddocks out to 45-60 days, they can shorten recovery for non-diverse pastures, especially ryegrass stands, to 30 days to keep quality. Recovery is determined by green leaves on tillers - six for cocksfoot and four for ryegrass and prairie grass.

Furthermore, cows strip-graze deferred pasture to keep up fibre. Even milking OAD, they're still shifting cows three or four times daily. Now cows go into pasture at 4,500kgDM and not lose production. They've also discovered deferred pasture of 170 days is best for drying off cows. Despite ME at 8.8, all mineral nutrient levels are similar to ME at 12 and all cows are content.

Reckons he grows only 8-9 tonne DM/yr, so not a high pasture performer and yet while DairyNZ says cows need 12.7kgDM to produce one kgMS, Adrian has done it with 9.8kgsDM which DairyNZ calculated while a monitor farm for four years – Adrian Frei.

Increasing dietary carbohydrates delivers MUN in the mid-teens over the season but 7-10 in spring coinciding with 7% empties. Before these changes, empties reached 14%. They get 83% in-calf at six weeks and only treated 12 cows for mastitis last year - they're used for calf milk, then culled.

In lengthening pasture recoveries beyond 37-38 days, they've observed swards achieving massive growth rates for 10 days of around 120kgDM/day. This is Andre Voisin's blaze of growth which most modern dairy farmers on a 21-day round never cash in on.

For pasture renewal, Russell will disc grass in and power harrow with no sprays. According to his grandfather's 1927-28 diary, they're using the same seed mixes today. Undersowing with a perennial diverse mix of grasses, legumes and forbs into existing ryegrass paddocks when dry is his most successful strategy. Inoculating all bare seeds with vesicular-arbuscular mycorrhizae (VAM) fungi works so well that five-day-old seedlings have root rhizomes covered in soil. That's immunity.

They haven't applied NPK for nine years as plenty is available in topsoil. Instead, they apply humates, lime flour, manganese, Ocean Base and Biosea products. Russell puts a slurry together, delivering lime at 50kg/ha per application, boron 200g, zinc 200g, humates 3kg/ha, magnesium sulphate 25kg/ha and mussel shell 5kg/ha, designed to stimulate soil biology. Originally, he was doing three applications per year and is now at two but wants to get down to zero.

As a result, they're observing topsoil deepening and darkening while water infiltration and holding capacity improves. A simple trick to transfer and target fertility involves Russell coaxing cows to particular spots in paddocks and stopping. While waiting they'll drop dung before he moves them on. The sound of dung hitting pasture is the beat the Healds march to. You can't do this if you're chasing production.

In meeting Russell and Charlotte, nothing appears a drama. They simply enjoy being curious and asking questions. They love their open days and showing people what they are doing because having other farmers visit their farm is how they get their best ideas.

Eco-profit Observations

- DairyNZ groups were no support during time of change.
- Buying a Tow and Fert got Russell into making slurries. That stirred his curiosity to ask more about soils.
- By keeping feeding simple, their farm working expenses average $4.50/kgMS including wages, at a time when DairyNZ's national farm working expenses are averaging $5.84.

- Milking once-a-day reduces livestock stress and allows a longer day to get other things done. It allows staffer Sahra to breakfast with her children.
- Simply lifted profit by investing in a business coach.
- Moving to polled Jerseys to cash up on increasing butterfat price.
- Dietary focus to lift carbohydrates and fibre to lift butterfat production.
- Sowing multi-species pastures delivers multiple benefits as they can either graze, cut or defer multi-species for later, meaning they graze ryegrass swards at 30 days to keep quality.
- Recovery is determined by green leaves on tillers - six for cocksfoot, four for ryegrass and prairie grass.
- Cows go into pasture at 4,500kgDM and don't lose production.
- Increasing dietary carbohydrates delivers mid-teens MUN over the season but 7-10 in spring coinciding with 7% empties.
- Pasture recoveries beyond 37-38 days; swards achieve massive growth rates for 10 days of around 120kgDM/day.
- Inoculating all bare seeds with vesicular-arbuscular mycorrhizae (VAM) fungi works so well that five-day-old seedlings have root rhizomes covered in soil.
- No NPK for nine years as plenty available in topsoil.
- Russell targets fertility by coaxing cows to particular spots in paddocks and stopping where they'll drop dung before he moves them on.

66. Branding Meat Quality

"We've learned over the decades of selling meat and poultry to upscale restaurants that our meat cooks 15% faster than what's in the store. We believe it's because our animals never secrete cortisol from being stressed, instead they secrete dopamine because they're happy. That brings up the question, can you eat happiness?"
Joel Salatin. Emily Penton Inner Clarity System podcast August 2025.

When marketing experts preach "branding," most people hear "spin", according to Harry Beckworth in The Invisible Touch. Broadcasting a slick, polished, unique or clever message isn't what makes a brand. Instead, its delivering the truth in an understated way while living an extraordinary promise from an ordinary message. It's a warranty, a vow, that a service or product will live up to its name and perform. Branding is powerful but it requires vision and dedication many farmers struggle with. That's why they abdicate marketing to someone else.

Promoting products falls into psychology, a world of smoke and mirrors most farmers have little time for while dealing with the day-to-day realities of running a farm. However, marketing starts inside farm gates and a large problem for New Zealand is that 80% of its meat production is non-brandable.

If you don't believe marketing starts with product, an interview in 2025 by Acres USA owner Taylor Henry lays the realities bare. North American vet Will Winter describes his experiences marketing grass-fed beef in US supermarkets. When they initially approached customers, they sold grass-fed mince at $4/lb versus feedlot (grain) finished at $1.99/lb. Nobody would buy it because of price, so they changed their marketing.

Firstly, they emphasised health benefits through CLAs, omega fat, etc. Didn't sell. Then they focused on environment outcomes such as carbon positive, etc. Didn't sell. Then on humane grounds, less handling etc. Didn't sell. It's clear, what marketing surveys say: Consumers' value is tempered by economic reality.

They only thing that worked was dressing up as cowboys and offering free products to shoppers in-store. Flavour and tenderness sold grass-fed beef. Only by creating a personal experience did customers change.

In the same interview, rancher Steve Campbell explained what quality meat is. In one testing using 12 ranchers' steaks where all rated equal for tenderness, the best flavour came from animals fully mineralised and their toxins removed. Minerals improve colour and flavour.

Also, mineral density stimulates satiety in humans. Instead of wolfing down a 32-ounce grain-fed feedlot steak, they tanked on a 10-12 ounce grass-fed steak.

Compare that experience to New Zealand farmer attitudes toward meat production. This is where Warren Buffett's quote of farmers rolling an ostrich egg into the hen house and saying, "Ladies this is what the neighbours are doing!" comes home to roost. The industry will always be about volume and weight because that is how industry agencies are funded. If farmers aren't prepared to serve cubes of succulent meat to indifferent shoppers, they lose that direct connection.

Just as Buffett and Bogle observe, professionals offering complicated solutions crank up their fees to profit from investor ignorance, impatience and greed – Consultants Breed Costs.

Allan Nation looked for lessons in the wine industry for meat to follow. As Californian wine volumes grew, prices reflected quality. While quality was relatively easy to discern, creating it was another story, because those techniques were secret. Many thought they involved post-harvesting technologies, such as French instead of American oak casks, storage temperatures and stainless-steel vats for mixing. But they found you can't "buy" quality.

It's the same in the meat industry. Ageing meat was thought to turn poor quality meat into something exceptional, but ageing just intensifies flavours. Any off-flavours grew in potency. Tenderness doesn't alter much after ageing five days and the lack of continuous fat cover exposes carcasses to bacterial infections. Fat is a key driver in texture and flavour as even minced meat must have 20% fat to taste exceptional. And it's better still if fortified with offal.

Vintners also point out soil qualities influence grape flavours and so do meat producers, as proven earlier. An algal, fishy flavour in grass-fed meat can be traced to acidic soils with low organic matter and poor mineral values. Drier climates with less leaching tend to produce better flavours. The USA's western grazing lands tend to produce the best meat, though other specific areas with highly available calcium and phosphate can compete.

Wineries say grape varieties influence flavour, while the meat industry endorses particular breeds. In its 16-years the New Zealand Steak of Origin competition became a bonanza for Angus. But any grape can excel for flavour and its best to find livestock that suits your weather and environment. All traditional livestock breeds started out as adaptations to local climate and soil.

A 2023 Beef and Lamb NZ study tour heard an interesting observation from US feedlots. When Angus and Hereford steers arrive at the same liveweight and go through to the same carcass weight, Herefords arrive at that a month later but ate less feed to get there. Who's looking at feed conversion efficiencies and meat quality?

So while there are many things that go into making a premium product to create a brand, there is a paradox. When buyers of quality wines frequent smaller boutique operations they're looking for exclusivity. That's what's prevented large American corporations cornering the specialised wine market. Consumer perceptions between quantity and quality determine the limits of niche size.

As Nation wrote in his Al's Obs book in 2012, it's the same with meat. There's little advantage in being a large corporation in the American grass-fed meat business because efficiencies of scale are restrained. Going back to Henry's interview, if customers are satisfied with smaller, more nutritious portions, how committed is the New Zealand meat industry to linking taste, nutrition and serving size?

> In the last 50 years, New Zealand annual farm income has only outpaced annual farm costs twice – Debt Creates Futures.

With the current state of supply and meat processing in New Zealand, what market signals do farmers need to create a branded product, or is that simply impossible because of the country's variation in climate, soil type and farmer skill and attitude?

Eco-profit Observations

- A brand can be defined as a warranty, a vow, that a service or product will live up to its name and perform.
- A marketing problem is that 80% of New Zealand meat production is non-brandable.
- If price creates customer resistance, production claims do not shift product. Only a personal experience with those who grow it challenge consumer hesitancy about paying a higher price.

- Meat with good mineral composition improves taste and triggers satiety, reducing serving sizes.
- American wine industry found it couldn't buy quality.
- Ageing doesn't fix poor-quality meat. It'll intensify off-flavours. No benefit to tenderness after five days' ageing. Fat is a key driver in texture and flavour.
- Drier climates with less leaching tend to produce better meat flavours, but other specific areas with highly available calcium and phosphate can compete.
- Any grape can excel for flavour and it's the same with livestock. It's best to find livestock that suits your weather and environment.
- Herefords arrive at a targeted feedlot carcass weight a month later than Angus but ate less feed to get there. Who's looking at feed efficiency?
- Buyers of quality wines frequent smaller boutique operations because they're looking for exclusivity.
- There's little advantage in being a large corporation in the American grass-fed meat business because efficiencies of scale are restrained.
- Can New Zealand create a branded meat product, or is that simply impossible because of the country's variation in climate, soil type and farmer skill and attitude?

67. Bill and Sue Garland - Supply Chain Strategy

How many New Zealand farmers have empathy for supply chains beyond their own gate? As many are unwilling or unable to invest time with red meat consumers, this handicaps industry progress. But there are commodity producers willing to work with processors and marketers to lift red meat's game.

Bill and Sue Garland are one such couple. Their 430ha sheep and beef property is snug against 8km of predator-proof fencing on the slopes of Mount Maungatautari, Waikato. While Bill's family has been here three generations, diligent research and pre-historic artifacts confirm human presence on a north-south trading route before Māori decided to stay. Māori settled here because it was above the frost line and had friable soils for kumara.

A chat with his father in mid-1970s always stuck with Bill. They were overlooking a 17ha uncleared hill block when his father said he wasn't developing it. They already had enough land to manage and it had good water which would be valuable in the future. He didn't trust his sons to respect his wishes so he registered a QEII covenant, the 11th block in New Zealand. What's significant is while the block had little grazing value, felling its tawa could have wiped their debt.

Some 15 years later Bill was on Federated Farmers board and invited to go globetrotting. That experience deepened his awareness of supply chains. With UK farmers grappling with new environment rules, he thought it wouldn't be long before markets demanded the same from farmers here. He realised New Zealand had to supply high quality products because volume was so small on the world stage. That meant committing to supply agreements.

They met John Atkins selling New Zealand lamb at a Whole Foods supermarket in California. Atkins basically set up the New Zealand meat trade there and his commitment to breaking into a select market inspired them to supply Whole Foods through Te Kuiti Meats.

To reward their skill as accurate suppliers, Te Kuiti Meats twice sponsored the Garlands to visit Whole Foods. In return they hosted their staff, plus New Zealand owners of the North American distribution company. In mixing with supply chain personnel, they learned first-hand what the systems and customers needed.

Visiting overseas grocery chains made Bill aware of many other supply chain links, from marketers to those selling on shop floors. When meat is fresh not frozen, it ages in a gas flush for 18 weeks, tenderising it and maintaining a fresh red colour. Once opened it has a four-day shelf-life alongside fresh lamb. If Whole Foods can't sell meat in four days, it becomes kebabs. If kebabs don't sell, then sausages. Each breakdown represents less premium. Products must have quality to provide more options.

Meeting customers in overseas supermarkets is a great way to deepen public awareness of the meat trade as many are misinformed about livestock and their relationships with farmers and the environment. A woman approached him stating she would never eat New Zealand meat because our farmers skinned lambs to mother-on orphaned ones. She hadn't realised only dead lambs are skinned.

On his travels Bill learned consumers trust family farms far more than corporations but he questions if farmers take advantage of this. With supply contracts, how many are aware of meeting their obligations and the industry consequences if they don't? For example, if farmers hold lambs back for heavier weights when premiums drop, they risk the whole market because buyers start looking elsewhere. It's said 80% of New Zealand meat is non-brandable. How much of that is due to farmers unwilling to commit to top quality, delivered on time, every time?

For farmers, supply contracts are a double-edged sword. Firstly, supplying means pushing animals out the gate. Bill's commitment to contracts increased because he knows who's buying his product. He has a relationship with his buyer based on trust. Secondly, and more important than premiums, it guarantees killing space.

> However, while money was good, start-up costs pushed through to year two when it became clear lamb trading profits weren't covering pastoral development costs – Michael and Catherine Reilly.

However, it also requires keeping meticulous records to ensure certainty and traceability for supply chain management. Now Garlands are NZFAP and GAP accredited for their meat.

They've also become great advocates for landscape plans. Not only do they document land management practices, landscape planning improves farm efficiencies by identifying locations of fences, shade, shelter and clean water. With infrastructure, for example, Bill moved the shearing shed to the centre of his farm and reduced mustering times.

> A 50c pine seedling in 1987 becomes a $150 tree 25 years later with few expenses beyond sunlight, CO2, water and time. Essentially, forestry allowed Mark to pay-off Waipari within his working career as well as creating options for carbon credits and family succession – Mark Warren.

Landscape planning software helps prioritise land use. For example, a very steep gorse block (500ft to 1,400ft) of 52ha, with 8ha in bush, was bulldozed, burnt and resown by hand. They hand-cleared the steepest 15ha from gorse and barberry and planted forestry for harvesting by cable winch to the top. The rest was subdivided into five paddocks, cultivated, re-grassed and established with a water system from a ram pump delivering 20,000litres/day. That investment lifted production from 7su/ha to 12su/ha. It also reduced muster times from two hours to 20 minutes, and meant less time applying fertiliser applications and on controlling weeds like gorse.

However, the biggest drawback is many industry-backed planning procedures produce static images that sidestep future farm resilience. When linked to time-constrained consents it prevents continuity, reducing certainty for business development.

There's always more to learn in the red meat game. After 25 years re-grassing, Bill and Sue discovered deferred grazing has no spraying or cultivation costs, produces better results than young grass for 8-9 months and reduces Californian thistles. Their records show production increased from nine ewes and 18 lambs/ha to 12 ewes and 24 lambs/ha. Deferred grazing inspires them to revisit soils, water and biodiversity.

Eco-profit Observations

- Father valued landscapes beyond their income generating potential. Put 17ha in QEII to ensure sons would do the same.
- Releasing New Zealand's small meat volumes on the world stage will require committing to contracts supplying high quality products.
- Travel overseas as reward for being the most accurate supplier allowed them to investigate what supply contracts meant to customers.

- Products must have quality to provide more options for lengthening product shelf-life.
- Some 80% of New Zealand meat is non-brandable. If farmers hold lambs back for heavier weights when premiums drop, they risk the whole market because buyers start looking elsewhere.
- Supply contracts mean committing to pushing animals out the gate but they also guarantee killing space.
- By documenting land management practices, landscape planning improves farm efficiencies.
- Biggest drawback is many industry-backed planning procedures produce static images that sidestep future farm resilience.
- Deferred grazing has no spraying or cultivation costs, produces better results than young grass for 8-9 months, reduces Californian thistles and lifts production.

68. Colin and Keryn Brown - Meating Consumer Expectations

When your multi-million-dollar technology business collapses how do you start again from scratch? Colin and Keryn Brown decided to grow and sell the world's best meat.

They run a little 26ha property with a big reputation. When their computer business folded they settled at Lake Karapiro, near Cambridge. Colin always had a love for Simmental cattle, and this became the embryo of their next business. But in a location prohibiting synthetic fertilisers and without new pastures, how did they produce award-winning meat?

Lambing performance dropped from 148% to 110% two years later. He admits overgrazing too many paddocks and spending way too much on superphosphate creating in a copper deficiency resulting in ewes losing condition. Phil and Viv Gray.

Successful entrepreneurs know to focus on customer experience and work backwards. Colin, impressed by Seth Godin's Purple Cow, knew he had to go beyond a black and white product. Jim Collins' books Built to Last and Good to Great inspired Colin to ask, "Why can't I produce the best meat in the world?" With typical zest, he made a start. But when he was pushing meat packs out the door he was already asking whether Simmental was the best meat?

He knew good meat quality was embedded in genetics. The Browns won their first Steak of Origin in 2009 with a Piedmontese, yet despite this breed having a tenderness gene it has no fat, so no marbling. Ultimately, fat equals flavour.

Intramuscular fat (marbling) has a huge role in meat-eating quality. Colin crossed with Angus and Wagyu, looking for the right mix to lift marbling, then realised after five years that the genetic potential was limited. So, he started again with registered cows, but finding any was challenging. Luckily Lawson Angus (Australia) sent over embryos. While this added time to their plan, the sweetener was they only paid for females. Bulls went to the South Island.

Genetically, they aim for calving ease, high growth and plenty of marbling to calve progeny in Australasia's top 1%. But keeping older breeding cows is still important to Colin. They test for eye muscle and marbling score, so only breed from these attributes. That increases the potential for better eating quality.

Selling meat directly to consumers changes the incentive to make things happen, commodities shift focus elsewhere. Colin notes the United States has three categories for buying steak, Prime, Choice or Select. Select has the least fat and has almost disappeared because people won't buy it. New Zealand processors don't pay a premium, so who cashes in on consumers paying more for steaks that shrink?

Sirloin, scotch and eye fillets are separated out for selling directly to consumers as beef packs. Then, what to do with cheaper cuts? They tried many meat products, but the unfair advantage of pie-making is the pie takes more parts of the animal. Two beasts make up 3000 pies. One professional did all the flavour testing while Keryn came up with the recipes; another unfair advantage Colin proudly mentions.

Building a market was key. They contacted everyone they could think of to take pies, then got on Facebook. The response was huge with 10 new customers a day, generating way too many orders. That's a problem every business wants. Their website is run from India and updates overnight.

They built their own app to trace every sale. They found 50% of buyers only buy once. Though they found this slightly disappointing, online marketing gurus advised them this was huge. Furthermore, what these gurus found unbelievable was that the rest buy 7.7 times, meaning on average, they can expect each customer to make 3.85 purchases from Lake Farm Pies.

Pie processing starts with slaughtering animals in Auckland, butchering in Hamilton and baking the final product at Mount Maunganui. Every second Tuesday pies are fresh chilled and couriered around New Zealand for the same cost, with the North Island subsidising the South Island. The Browns start with a 2am alarm, are on the road to Tauranga by 3am and packing by 5.30am at Mount Maunganui to be out by midday.

Each shipper contains eight pies in a custom cooler bag plus a 1kg chill pad. Pies wrapped in greaseproof bags stay colder longer than cardboard boxes. They discovered this while proving their product was perishable after 48 hours to get courier status during Covid. By using previous career contacts, they source packaging cheaper and faster from China than New Zealand. As a non-public business Covid was easy to handle.

Colin finds most stud and commercial breeders' "who cares" attitude to well-marbled meat disappointing. His aim is for a brand to compete with the likes of Wagyu, with well-marbled beef from animals that finish a year earlier than Wagyu. That standard largely uses EBVs or high ultra-sound values.

The experience introduced him to possibilities where both parties can benefit without a sense of being used, a business insight reinforced when meeting Joel Salatin years later – Dennis Nieuwkoop.

He proposed naming this standard "Ahurei project," a Māori word meaning "special and unique", to grow beef for tomorrow's world. Although he has lost the desire to evangelise this to growers, it has not diminished his own breeding goals.

In 2025, they'll shift from fortnightly to monthly deliveries. This change will coincide with more heifers going on the open market to earn a premium for their genetic marbling and growth potential.

After the trauma of losing their computer business, they've set up Lake Farm Pies to be easily shut down; a savvy exit strategy for their retirement years.

The Browns demonstrate nothing beats passion. After all, they got down to the last 20 steaks in the Steak of Origin cook-off seven times in nine years, and won it twice. Being a smaller operation has advantages but the greatest gains come from focusing on customer experience.

Eco-profit Observations

- In a location prohibiting synthetic fertilisers and without new pastures, how did they produce award-winning meat?
- Successful entrepreneurs know to focus on customer experience and work backwards.
- Fat equals flavour. Piedmontese has a tenderness gene but no fat.
- Genetically they aim for calving ease, high growth and plenty of marbling.
- In the US, Select grade has the least fat and is disappearing because people won't buy it.
- Unfair advantage of pies is they take more parts of the animal.
- Facebook generated significant orders.
- Found 50% of buyers only buy once but customers overall make an average of 3.85 purchases.

- Pies wrapped in greaseproof bags stay colder longer than cardboard boxes.
- Created a marbling standard for well-marbled beef from animals that finish a year earlier than Wagyu.
- They've set up their business to shut it down easily; a savvy exit strategy for their retirement years.

69. Greg Crombie - Branding the Best

The best examples of marketing in agriculture are usually found in the wine industry. While marketing lays the strategic path businesses take, branding is the emotional efflorescence in its wake – the effect the product or service leaves behind. Branding isn't just a logo.

Marlborough's Leefield Station is owned by the Marris family, entrepreneurs who look at market gaps and reverse-engineer ways to fill them. They are brand builders, developing Wither Hills and The Ned labels, very successful local wines. In 2012 they added Leefield Station, a farm tracing its history to 1840s. Today, 2,200ha of hill and river flats is roughly split into thirds with vines, pastoral hill country and QEII covenants and scrub country.

While organic is restrictive, having to work around issues simply means being organised and communicating better – Willie

Greg Crombie runs the pastoral side of the business - the breeding and grazing land not planted in grapes. Since arriving in 2018, Greg has overseen a lift in stock numbers from 3,000su to 5,000su in six seasons. He spent many years managing operations that pride themselves on production, but Leefield Station's branding brings a new dimension to farming. Livestock are secondary, they are part of the property's unkept image. It's why pines will never be planted. Other constraints include no high-analysis fertiliser to keep GHG emissions low for the winery.

That said, they're six years into a 10-year substantial pastoral development, including a stock water scheme gravity-feeding 18,000 litres/day, pumped from the Waihopai River up 180 metres using two solar-powered pumps through 63mm pipe. The development includes over 50km of pipe, 100 troughs plus another 50km of fencing. It'll take five years to get a return.

Solar panels are placed on plateaus out of sight so an unkept look is maintained. This development supersedes dams as diggers are too expensive to upgrade and maintain them. New stock tracks are designed to capture water and keep moisture on the hill.

Pastoral blocks are cut into 20ha paddocks through subdivision. This creates longer rotations by allowing ewes to push out to 35 days recovery and lift calf weaning weights by up to 30kg. With grazing, Greg operates an 80:20 rule -

consume 80% and leave 20% over June-November in case spring rains don't come.

There's 2,200 Romney ewes squeezing out 150% lambing. Some 5,500 lambs are grazed under vines during winter, including any trading lambs. This is the only time animals are on the flat, except when they come down for shearing. Some 90% of the buildings have appeared in the last six years, all designed to reflect building styles of the late 1880s.

As many as 270 Angus cows graze the steepest faces, carrying nutrients to hilltops where many new troughs are sited. In doing most of the work, these animals prevent native bush covering hillsides while younger livestock classes graze easier country. With all flat country in grapes, lucerne balage is bought rather than made as a supplement or deferred grazing. Calving is finished by the end of September, calves are weaned in February and sold at Feilding for good premiums in spring.

Soils at higher altitudes test very well, so Greg focuses on adding more nutrients to lower country, mainly elemental sulphur and phosphate. Fertiliser requirements are calculated by removing cows so applications target sheep production - cows are to clean up tag. Topdressing planes use variable rate applicators, meaning as country gets steeper, less falls out of the hopper. This results in plenty of sub-clover on hills.

All these practices reflect how Leefield Station strives to live up to its brand. They challenge a marketplace perception that customers live by brand values, that nobody cares where product comes from. Meat companies approaching Leefield Station to collaborate haven't got that point yet and why double branding doesn't work. Getting behind a leading brand with a powerful singular vision has many benefits.

The board wanted a pathway to align new staff with these objectives while creating a staff culture of continual improvement for the Glenaray "brand" to push its capacity – Simon Lee.

Leefield Station's marketing message can be seen in several smaller livestock enterprises capturing specific markets. These are not centrepiece enterprises but all are designed to demonstrate how livestock and environment benefit each other.

They have 140 registered and 120 commercial Angus cows and sell 18 bulls a year. These animals graze in the commercial mob, not on their own, so they're hardy from climbing hills. All cattle replacements come from this stud herd, the rest are sold.

Greg's exploring new lamb production with 100 modern Hampshire ewes delivering lambs 1-2% heavier than other breeds he's trialled. As they grow and get weighty, they marble. They're more resilient than other breeds and don't need rain to keep growing after Christmas when feed hardens off. Greg sold seven rams a couple of years ago and 14 last year so there is a potential market for them too.

Hampshire lambs and Berkshire pigs are slaughtered for restaurants, VIPs and guests. Berkshire pigs are kept in a one-hectare walnut grove planted in 1940. They're ideal for prosciutto cuts because the intramuscular fat in the pork adds moisture, which is great for ham.

For six years, Leefield Station has been producing jerseys, hats and blankets from 28-micron Romney wool. These are great items to compliment wine gift-packages or sold on their own and demonstrates the pride the brand has for its own livestock and how they groom hill country landscapes.

With no low-hanging fruit in agricultural commodities, branding is one way for industry to get more value. It requires new skills to find and leverage an attribute, then seek a customer base who resonate with it and sell to them. As Greg mentions, even if you're adding only 30% to the price, you're still moving forward.

Eco-profit Observations

- Livestock are secondary; they develop the property's unkept image. It's why pines will never be planted.
- No high-analysis fertiliser to keep GHG emissions low for the winery.
- Solar panels for pumping stock water are placed on hillside plateaus out of sight so an unkept look is maintained.
- New stock tracks are designed to capture water and keep moisture on hill.
- Longer rotations by allowing ewes to push out to 35 days recovery and lift calf weaning weights by up to 30kg.
- Some 5,500 lambs are grazed under vines during winter.
- Angus cows graze the steepest faces, carrying nutrients to hill tops where many new troughs are sited.

- Fertiliser requirements are calculated by removing cows so applications target sheep production, cows are to clean up tag. Topdressing planes use variable rate applicators.
- Meat companies approaching Leefield Station to collaborate don't understand why double-branding doesn't work.
- Hampshire lambs are more resilient than other breeds and don't need rain to keep growing when feed hardens off after Christmas.
- Berkshire pigs are kept in a one-hectare walnut grove planted in 1940. Their ham is ideal for prosciutto cuts.
- Sell jerseys, hats and blankets from 28-micron from own Romney wool.

70. Jamie and Nicky Gaddum and Robert Haselsteiner - Branding Kiwi Culture

When Austrian Robert Haselsteiner bought Matangi, a 480ha hill country property near Havelock North, Hawkes Bay, he wasn't interested in traditional New Zealand farming, but Angus cattle got him excited.

He challenges everything about farm production, which manager Jamie Gaddum finds refreshing. His desire to improve New Zealand steak quality launched them into a creative crusade - why not grow the best grass-fed Angus beef in New Zealand for New Zealanders? So emerged Matangi Angus Beef.

With a background in merchant banking, Robert has no experience with meat, but knows New Zealand is the home of grass-fed healthy meat. And winning six gold medals plus Supreme Champion at the NZ Outstanding Food Producer Awards (2025) shows the industry thinks they have potential. What's truly challenging to them is lifting meat's value to New Zealanders.

European culture focuses on local; around every corner is a different cuisine because diversity is key to commercial survival. Every place specialises in cooking animal parts differently. Why create a branded product to build relationships 18,000km away? It's easier transplanting European cuisine culture here and valuing locals and tourists.

The first thing was better genetics. Intramuscular fats drives Jamie's breeding programme as IMF has 80% heritability. Cattle studs don't give up registered cows but he found 28 in-calf with desired carcass traits. DNA genomic testing determined which female calves were kept for replacements, the rest was sent to high profile chefs to discover if they were on the right track. Rissington Cattle Co genetics is the backbone of the Matangi herd.

Next was livestock management. Jamie fully feeds livestock to accentuate marbling, it is the last fat to gain and first to drop off. Animals bloom on grass all year round benefiting from high covers, balage and hay, all home-grown, no non-grass supplements. Each mob is kept together for better social cohesion and less fighting. Low stress is essential for tenderness, so no huntaways are used, only heading dogs, to reduce stress and keep meat pH low.

With Matangi as the breeding block, Jamie, and his wife Nicky bought Atua, a 390ha property at Elsthorpe, for growing young stock and finishing cattle over summer. Flat to rolling country, it grows grass in summer when Matangi dries out.

Employing Mike Sugden at Matangi and Mark Murphy at Atua lifted carcass efficiencies through grazing management. Reserve grades grew from 49% to over 80% in three seasons with 65% of carcasses reaching marbling scores of five or higher. People pay more for this grade, which produces more income.

While culling their sheep flock buffers a shrinking feed supply, especially through winter, offering lamb to customers was an opportunity worth pursuing. Ewes scan 180% and lambs average 19.4kg cwt weaned, but 21.4kg through the shop. They run two lambing dates to supply 15 lambs per week for seven months. Lamb IMF is measured at the abattoir with Meat Eating Quality (MEQ) averaging 3.36%. They cross Romney ewes with Kelso Prime Q rams for tenderness. Globally, lamb eating quality is in its infancy.

Seasonality makes consistent meat-eating qualities a challenge but spilt calving smooths out supply over a 10-month season. With cattle, nothing is killed before 24 months, meaning a higher cost of carry but worth it. Mobs are weighed often but selection is on condition.

> The Shorts believe their greatest insights for dairy production came from learning soil function. Improving soil generates healthier animals, meaning they're not worrying about treating animals for problems – Kenneth and Rachel Short.

They slaughter four or five beasts every Friday at Progressive Meats. Carcasses are hung for three days, then independently graded for meat colour, fat colour, pH and marbling score to determine whether it has reached reserve grade. Carcasses are picked up Mondays as quarters and hung in their butchery for 21 days. Heifers score better than steers, always producing smaller cuts and have more fat and greater consistency reaching reserve grade.

The key to starting businesses is low overheads. Instead of hiring a premises, they contracted processing on a per-kilogram basis, meaning throughput determined costs. Their butcher worked out of his own shop and he grew with them. The only problem was the more they grew, the more it cost.

When cyclone Gabrielle flooded the butchery in February 2023 they lost all branded revenue for several months. So they rented a Hastings building and within 12 months developed their own processing facilities. Now customers buy cuts at their retail butchery, which has a showroom demonstrating brand values.

Nicky runs the butchery, which employs two butchers. It includes a dry-aging cabinet customers can look through. Dry-aging occurs after hanging 21 days. The dry-aging cabinet is maintained at 80% humidity and two degrees. During dry-aging, enzymes break down connective tissues, tenderising the meat and intensifying flavour. Shrinkage is 8%. Clients leave purchases to dry-age for up 100 days, carving off what they want when needed.

They liked the idea of controlling the whole supply chain from woolshed to wardrobe. They're cashflowing it themselves and learning how to develop a brand during tough economic times – Paul and Prue Ensor.

Marketing concentrates on social media, TV, videos and aligning with other brands. Word of mouth is best so chefs are ideal for introducing potential customers to Matangi Angus Beef with Matangi's logo on their menus. Nothing is better than customers saying, "Ate your meat at such and such and have come to buy it."

Chefs have ordered online for weekly delivery since 2019. Boxes are delivered to doorsteps so focus is packaging and presentation to ensure meat arrives in perfect condition and looking its best. It's easy to sell high-value cuts, but low-value cuts are more challenging, so this is where opportunities lie. Meat connoisseurs provide insights about lesser-known cuts such as Aitchbone, Bavette and Picanha. There's no discounting, so meat is never on special. They're always looking for another market.

Businesses rely on one thing - loyalty of customers. Nothing else comes close to determining real value. The team at Matangi Angus Beef knows European culture has no shortcuts. Top product and service is the only way to build a brand.

Eco-profit Observations

- European culture focuses on local - around every corner is a different cuisine because diversity is key to commercial survival.
- IMF has 80% heritability so this drives their breeding programme.

- Fully feeds livestock to accentuate marbling, the last fat to gain and first to drop off.
- No non-grass supplements, low stress is essential for tenderness, so no huntaways to keep meat pH low.
- Grazing management lifted carcass efficiencies with 65% reaching marbling scores of five or higher over three seasons.
- They run two lambing dates to supply 15 lambs per week for seven months. Globally, lamb eating quality is in its infancy.
- Spilt calving smooths out supply over a 10-month season.
- Heifers score better than steers, always producing smaller cuts and have more fat and greater consistency reaching reserve grade.
- Instead of hiring a premises, they contracted processing on a per-kilogram basis meaning throughput determined costs.
- Dry-aging occurs after hanging 21 days. The dry-aging cabinet is maintained at 80% humidity and two degrees.
- It's easy to sell high-value cuts, but low-value cuts are more challenging, so this is where opportunities lie. There's no discounting, so meat is never on special.

71. Why Bear Grylls should Spearhead Red Meat.

Better to pay the butcher than the doctor.
Irish proverb

The Irish keep buying our best rugby players and coaches. Surely we can scrape up a few million to pinch their single best celebrity to promote red meat.

Bear Grylls understands both global and human health. He knows pasture is great for saving the planet and that the livestock grazing it are destined for the dining table. He knows what's wrong with the food pyramid and lives life accordingly. This man has a commitment to red meat and would be an awesome ambassador.

Name another celebrity travelling and interacting more with natural ecosystems than Grylls. He knows the role of predators is to bunch herbivores and keep them moving across grasslands, ensuring herds only breed from the strong. Their disturbance cycles carbon into soils to absorb moisture and prevent erosion. No other global celebrity championing the environment comes close to carbon positive because they cannot relate these principles to livestock farming. And few, if any, would copy his diet.

When Louis Theroux interviewed him in 2022, the whole world heard of his time being vegan and how embarrassed he was of the cookbook he wrote. Being vegan collapsed his energy and health, and increased his flatulence. It's vegan diets driving climate change and it's a shame he never cooped himself up in a methane chamber to find out how much greenhouse gases vegans emit.

He's got little time for grains, vegetables, poxy seed oils and corn syrup. This is a man who has eaten sheep's eyes, deer hearts and goat testicles and slurped up yak blood. He's gutted and slept in a camel carcass and worn seal skin to swim frigid waters. He knows animals inside out and has immense enthusiasm for the product. He even has his own range of ancestral supplements made of beef heart, liver and lung, all grass-fed.

While he's all for natural foods, lentils, pulses and mungbeans appeal as much as swallowing water squeezed from elephant dung. That's something else he doesn't recommend.

So, what flipped him? Watching his son's speedy recovery from various nutritional aliments caused by a vegan diet. It involved dropping refined processed carbs (flour and sugar) and vegetables, and replacing them with red meat, eggs, fruit, honey and salted butter and cooking with lard or tallow. He tried it himself, especially after being ravaged by early covid variants when he found that eating more vegetables didn't improve his recovery. In fact, he found it made him worse.

This is up there with feeding cows mashed brains and spinal cords - science where only farmers are held accountable – GMO Ryegrass only Treats Symptoms.

Guzzling down raw vegetable smoothies almost gave him kidney stones. That condition and its pain disappeared, along with skin issues, stomach problems and lethargy, once he began to eat red meat. He was always hungry on a vegan diet and maybe that's why the vegetable servings rule is five a day. Now he remains satisfied with little if any snacking between his two daily meals.

He's found a red meat diet is simply more fun and delicious. It's this appeal at the root of every child's sense of adventure, and as the UK's top scout he ought to know. He is clear about the health benefits from eating red meat - not just physical stamina and intellectual clarity, but emotional stability. With this energy to draw on he feels confident to explore such deeply personal subjects as life, survival and mental health.

Nation points out the only real price is what a beast gets at slaughter. Everything else is due to human greed and fear, all emotion – Market Cycles and Price Swings.

These experiences open the eyes of young people to ups and downs of living while learning to enjoy the present. As a parent himself, this is front and centre with his own children. Resilience is like a red meat muscle - it needs to be worked to get stronger. Many adults could do with this advice.

His TV programmes demonstrate adventure is deeply embedded in human DNA. Resilience involves tackling fears by relentlessly trying new ways of solving life's obstacles and boundaries. His favourite food is grass-fed beef because human fear releases a lot of energy and nutrients and grass-fed meats and fats have oodles of both.

And where will red meat fit into his life as he ages? Now 50, he's at the peak of endurance and health. His exercise regime focusing on strength and suppleness is mostly outdoors so red meat will continue to fuel that. Like many farmers, he knows being out in nature is a great healer. By the way, he prefers to walk barefoot and probably enjoys both feet in a fresh cow pat on a frosty morning.

Red meat changed Bear Grylls' life and New Zealand's livestock industry should celebrate a contract with him before Australia's MLA does. It's nutrition that will drive his success as he gets older. If he had the skills, he'd keep up with those shearing 250 80kg ewes a day.

In fact, why don't we put him in front of strong wool as well? I bet he's prayed for a woollen carpet and underlay to cushion a fall or two.

Eco-profit Observations

- No other global celebrity championing the environment comes close to carbon positive because they cannot relate ecology principles to livestock farming.
- Being vegan collapsed his energy, health and increased his flatulence.
- What flipped him to red meat? Watching his son's speedy recovery from various nutritional aliments caused by a vegan diet.
- Has his own range of ancestral supplements made of beef heart, liver and lung, all grass-fed.
- Now finds he remains satisfied with little if any snacking between his two daily meals, unlike on a vegan diet requiring five servings a day.
- With red meat energy to draw on he feels confident to explore such deeply personal subjects as life, survival and mental health, especially with young people.
- Like many farmers, he knows being out in nature is a great healer.
- At 50, his exercise regime focusing on strength and suppleness is mostly outdoors. He prefers to walk bare foot.
- Red meat changed Bear Grylls' life.

Adaptation: Why Bear Grylls should Spearhead Red Meat. Stuff November 18, 2024 and New Zealand Farmer November 5, 2024. p.28
https://www.stuff.co.nz/rural/360466864/why-bear-grylls-should-spearhead-red-meat

REFERENCES

Beckworth, H. (2000). The Invisible Touch; The Four Keys to Modern Marketing. Warner Books.

Bogle, J. C. (2017). The Little Book of Common Sense Investing. Wiley.

Brown, G. (2018). Dirt to Soil: One Family's Journey into Regenerative Agriculture. Chelsea Green.

Collins, J. C. Porras, J. I. (2004). Built to Last: Successful Habits of Visionary Companies. Harper Business.

Collins, J. (2001). Good to Great. Random House Business Books.

Godin, S. (2002). Purple Cow: Transform your Business by Being Remarkable. Portfolio.

Hocken, J. (2019). The Lean Dairy Farm. Wiley.

Marks, H. (2018). Mastering the Market Cycle: Getting the Odds on Your Side. Harper Business

Massy, C. (2017). Call of the Reed Warbler. Queensland University Press.

Nation, A. (2018). Creating a Family Business from Contemplation to Maturity. Green Park Press.

Nation, A. (2012). Al's Obs; 20 Questions and their Answers. Green Park Press.

Nation, A. (2003). Knowledge Rich Ranching. Green Park Press.

Nation, A. (1998). Pasture Profits with Stocker Cattle. Green Park Press

Popay, I. Champion, P. James, T. (2010). An Illustrated Guide of New Zealand Weeds, 3rd Ed. New Zealand Plant Protection Society.

Provenza, F. D. (2018). Nourishment: What Animals can Teach us about Rediscovering Our Nutritional Wisdom. Chelsea Green.

Savory. (2016). Holistic Management: A Commonsense Revolution to Restore our Environment. Island Press.

Sears, P. D. (1953). Effect of Day and Night Grazing on Annual Pasture Composition and Production. NZ Jou of Sci and Tech, Supplement 1, 68-77

Sears, P. D. (1962). Exploitation of High Production Pastures. Proceedings of the New Zealand Ecology Society.

Smallfield, P. W. (1970). The Grasslands Revolution of New Zealand. Hodder and Stoughton

Warren, M. (2018). Many a Muddy Morning. Harper Collins.

Common New Zealand Agricultural Terms

Block refers to a parcel of land. It could be an entire property or a number of paddocks within it. For example; hill block, river block, neighbouring block, back block, forestry block.

Cattle stop: cattle grid

Crayon: at mating, rams where harnesses with a coloured wax crayon to identify ewes that have been mated.

Dags refer to accumulating dung in wool around the backsides of sheep and lambs. Such animals are referred to as daggy and removing dags is known as dagging or crutching.

Dairy Farm system; New Zealand dairy farms are rated 1-5 based on how much feed they bring on the grazing platform. If nothing is imported it's considered a System 1, if 50% or more of feed consumed is bought into the operation, then it's a System 5 farm.

Fly/flystrike: when dags or wool become infested with fly maggots. Common practice is to dip or jet with insecticide.

Freezing works or works are livestock abattoirs. The term freezing works originates from New Zealand meat production being frozen for international shipping last century.

Glammies or Golden Lamb Awards was an annual competition run by Beef and LambNZ to identify farmers producing the best lamb eating qualities. It ran from 2007 – 2018.

Hoggets are rams, ewes and wethers less than one year old.

Micron: diameter of wool fibres. Ultra is less than 15, super fine 15-18.5, fine 18.5-19.5, average 19.5-24. These are used for clothing next to the skin. Coarse wool 24-35 or higher is for outdoor wear, insulation and carpets.

Milk Solids: New Zealand dairy cow production is calculated by milk solids. Essentially 15% of any volume of milk is made up of solids of which over half is fat and protein, the rest made of other compounds such as lactose and minerals.

R1 and R2 refer to young cattle, rising one year old (aged less than 12 months) and rising two year old (age less than 24 months).

Race is a long narrow chute usually prior to a loading ramp or drafting gates for animals to move through in single file. The drafting race is preceded by the working race where activities such as dagging and drenching occur. Prior to the working race is the forcing pen to funnel animals into the working race.

Ryegrass staggers: a condition where animals eat pasture infected with an endophyte producing a range of symptoms including losing the ability to walk or run when disturbed. Sometimes simply referred to as grass staggers.

Shearing is the removal of wool from sheep.

Station is usually a large farm, similar to the term ranch, often associated with extensive grazing grasslands. However, some properties keep the name station because they're the remnants of an original homestead with the rest sold off over generations.

Steak of Origin was an annual event run by Beef and LambNZ to identify farmers producing the best eating qualities for sirloin steak.

Stock Unit (su) represents a 55kg ewe raising a single lamb. As many farmers run ewes consistently heavier, 65kgs to raise twins, each ewe represents 1.3su. Beef cattle are considered 6su each and heavy yielding dairy cows are eight. The stock unit is used for base line comparisons throughout New Zealand to compare production and economic values.

Tupping is when ewes are joined by rams at mating time.

Two-tooths is the next age phase after hogget associated with the emergence of two incisor teeth on the lower jaw. As the animal gets older, more incisors emerge to replace milk teeth so that by the age of four years they gone through two-tooth, four tooth, six tooth and on to eight tooth or full mouth.

Wether is a castrated ram. These are usually killed as lambs. Some high country stations use wethers to graze challenging country which produces finer wool.

Yards means corral. It's where all the stock work is done. Sheep yards are often associated with woolsheds and many farms have the main working area covered to complete tasks in all weather. Sheep yards and cattle yards are different facilities specific to each class of livestock.

INDEX

C

D

E

F

G

H

I

J

K

L

M

N

O

P

Q

R

S

T

U

By the same author:

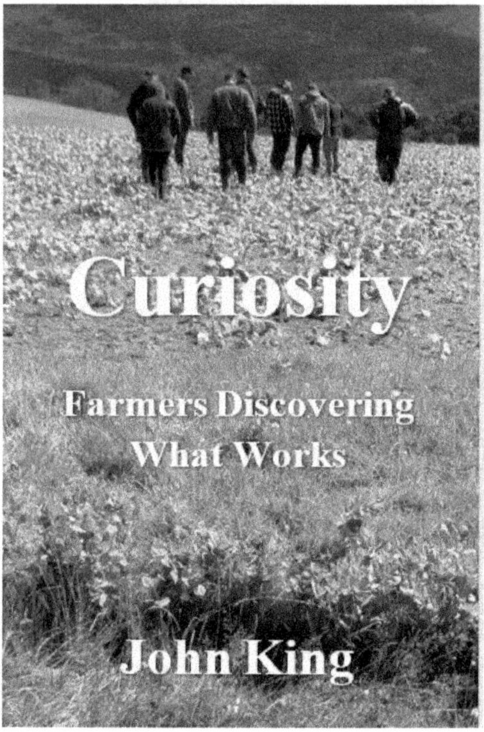

For those interested in what farmers have discovered that works for them….

- How farmer beliefs change.
- What multi-species pastures work.
- How do low-cost grazing operations become successful.
- Pioneering bale grazing in New Zealand.
- Strategy to develop a farm from an earthquake.
- Secrets to grazing large mobs of bulls.
- How quickly can soils change?
- The opportunity cost of planting winter crops.
- How averaging farm input costs creates pollution.
- How industry modelling costs farmers.
- How organic reduces financial risk.
- Why crop nutrition is key for success.
- Moving beyond a chemically dependent property.
- Power of hedgerows to lift property performance.
- Why farm experimentation is so important.
- The role of monitoring in success.
- High production isn't high profit.

And more…

www.ingramcontent.com/pod-product-compliance
Lightning Source LLC
Chambersburg PA
CBHW050522240926
51846CB00003B/660
* 9 7 8 1 9 9 1 1 8 8 7 6 2 *